Increasing Participatory Engage through Technology and Design

This broad-ranging edited collection highlights the role that technology can play in increasing community engagement in participatory design, now and in the future.

Challenging conventional notions of participatory design, this volume provides innovative solutions in areas such as democracy, sustainable living, smart cities, energy, and transport. Design has the capacity to engage citizens and communities in new and different ways, and since the onset of the Covid-19 pandemic, it has become increasingly important to rethink how services and engagement can be made digital. Each chapter demonstrates applied and potential uses of emerging technologies as a form of engagement with select case studies from South and North America, Europe, and Asia, all grounded in participatory theories, methods, and practices. Together, they explore how the green transition, the move to Industry 5.0, global-centric design, or systemic design approaches play a role in citizen engagement when current paradigms are being challenged. Leading practitioners and researchers cover topics such as crowd-sourcing information gathering, visual and systems thinking, creativity tools, participatory budgeting and governance, online participation, and citizen engagement to bridge communities and government.

This interdisciplinary book provides theoretical and practical knowledge for those who work with citizens, communities, and participants in multiple contexts, including designers in the fields of interaction and user experience design, product design, service and system design, social design, and design for policy and governance. It will also appeal to managers, public and private sector planners, engineers, and policymakers.

Mari Suoheimo is an Associate Professor at The Oslo School of Architecture and Design, Norway. Her main research focus is on how to approach wicked problems in service design. Suoheimo is also part of the editorial board of the *Design Journal* and the *Contexts Systemic Design Journal*.

Jonna Häkkilä is Professor for Industrial Design at the University of Lapland and a docent in Human Computer Interaction at the University of Oulu, Finland. She leads the user experience research group LUX focusing on design and technology and has received numerous grants, e.g., from the EU, Academy of Finland, and Interreg.

Ravi Mahamuni is Principal Scientist at TCS Research and Adjunct Professor at IDC School of Design, IIT Bombay. He is a Design Thinker with 29 plus years of experience; his research focuses on design for change, system design, and service design at scale. He believes in technology as an enabler while humanizing the services. He also advises several academic institutions on matters related to design, technology, and social change.

Design for Social Responsibility

Series Editor: Rachel Cooper

Social responsibility, in various disguises, has been a recurring theme in design for many years. Since the 1960s several more or less commercial approaches have evolved. In the 1970s designers were encouraged to abandon 'design for profit' in favour of a more compassionate approach inspired by Papanek. In the 1980s and 1990s profit and ethical issues were no longer considered mutually exclusive and more market-oriented concepts emerged, such as the 'green consumer' and ethical investment. The purchase of socially responsible, 'ethical' products and services has been stimulated by the dissemination of research into sustainability issues in consumer publications. Accessibility and inclusivity have also attracted a great deal of design interest and recently designers have turned to solving social and crime-related problems. Organisations supporting and funding such projects have recently included the NHS (research into design for patient safety); the Home Office has (design against crime); Engineering and Physical Sciences Research Council (design decision-making for urban sustainability).

Businesses are encouraged (and increasingly forced by legislation) to set their own socially responsible agendas that depend on design to be realised. Design decisions all have environmental, social and ethical impacts, so there is a pressing need to provide guidelines for designers and design students within an overarching framework that takes a holistic approach to socially responsible design. This edited series of guides is aimed at students of design, product development, architecture and marketing, and design and management professionals working in the sectors covered by each title. Each volume includes: The background and history of the topic, its significance in social and commercial contexts and trends in the field. Exemplar design case studies. Guidelines for the designer and advice on tools, techniques and resources available.

16. **Design for Education**
    Spaces and Tools for Learning
    Edited by Ana Rute Costa and Rachel Cooper

17. **Systemic Service Design**
    Edited by Mari Suoheimo, Peter Jones, Sheng-Hung Lee and Birger Sevaldson

18. **Increasing Participatory Engagement through Technology and Design**
    New Approaches for Participation
    Edited by Mari Suoheimo, Jonna Häkkilä, and Ravi Mahamuni

For more information about this series, please visit: www.routledge.com/Design-for-Social-Responsibility/book-series/DSR

Increasing Participatory Engagement through Technology and Design

New Approaches for Participation

Edited by Mari Suoheimo, Jonna Häkkilä, and Ravi Mahamuni

Routledge
Taylor & Francis Group
LONDON AND NEW YORK

Designed cover image: Mari Suoheimo

First published 2027
by Routledge
4 Park Square, Milton Park, Abingdon, Oxon OX14 4RN

and by Routledge
605 Third Avenue, New York, NY 10158

Routledge is an imprint of the Taylor & Francis Group, an informa business

British Library Cataloguing-in-Publication Data
A catalogue record for this book is available from the British Library

Library of Congress Cataloging-in-Publication Data
Names: Suoheimo, Mari, editor. | Häkkilä, Jonna, editor. | Mahamuni, Ravi, editor.
Title: Increasing participatory engagement through technology and design :
new approaches for participation / edited by Mari Suoheimo, Jonna Häkkilä and Ravi Mahamuni.
Description: Abingdon, Oxon ; New York, NY : Routledge, 2026. |
Series: Design for social responsibility |
Includes bibliographical references and index.
Identifiers: LCCN 2026006325 (print) | LCCN 2026006326 (ebook) |
ISBN 9781041130550 (hardback) | ISBN 9781041130543 (paperback) | ISBN 9781003667872 (ebook)
Subjects: LCSH: Technology–Sociological aspects. | Design–Social aspects. | Social participation.
Classification: LCC HM846 .I53 2026 (print) | LCC HM846 (ebook)
LC record available at https://lccn.loc.gov/2026006325
LC ebook record available at https://lccn.loc.gov/2026006326

ISBN: 9781041130550 (hbk)
ISBN: 9781041130543 (pbk)
ISBN: 9781003667872 (ebk)

DOI: 10.4324/9781003667872

Typeset in Times New Roman
by Newgen Publishing UK

Contents

Figures

Tables

Contributors

Rauni Äärelä-Vihriälä is a postdoctoral researcher in the REBOUND project (funded by the Strategic Research Council, STN) and a project manager of the CAP-SHARE project at the Faculty of Education, University of Lapland. Her research focuses on Sámi education, sustainable development, and biodiversity awareness in the Arctic and Sámi context, with an emphasis on equitable education from an intergenerational perspective in relation to climate change adaptation and biodiversity change.

Monica Aejmelaeus (ThM) is a doctoral researcher at University of Tampere and a Digital Pedagogy Specialist at Metropolia University of Applied Sciences, Finland. Her research interests include digital pedagogy, continuing education in the energy sector, and participatory research methods.

Yue Ang, Lecturer in Law at Brunel University London, researches company law, corporate governance, and sustainability, focusing on embedding sustainability in legal education. She employs innovative knowledge-sharing methods and teaches commercial, land, and tort law. Yue holds a PhD (Birmingham), LLM (UCL), and LLB (Westminster).

Anniina Autero is a head of development for the City of Tampere. She has a strong interdisciplinary background in safety, security, resilience, and sustainability, with expertise in applying digital twins and metaverse technologies to participatory urban planning. Anniina's work bridges research and practice, fostering international collaboration to enhance wellbeing and adaptive capacity in cities.

Conny Bakker, University of Technology Delft, The Netherlands. Bakker is Professor of Design for Sustainability and Circular Economy. Her research field is the development of design methods for circular and sustainable product-service systems. She explores strategies such as product lifetime extension, reuse, remanufacturing and recycling, and the business models that enable these strategies.

Ella Björn is a doctoral researcher at the University of Lapland in the Faculty of Art and Design. She works as a specialist in the Frostbit Software Lab of the Lapland University of Applied Sciences. Her research interest focuses on investigating embodied empathy in service design through creative technologies and tools.

Raúl Castaño-Rosa is senior researcher and co-chair of Sustainable Housing Design research group (ASUTUT) in Tampere University, leading research on socially resilient housing design and transformation. Raúl is also a visiting young professor at the Jiangsu Province Engineering Research Center of Urban Heat and Pollution Control, Southeast University (China) and Coordinator of the Finnish EPAH Antenna and member of the Scientific Advisory Board in the Energy Poverty Advisory Hub (EPAH).

Angelique C-Dina (PhD Student, Interdisciplinary Design & Media, CAMD, NU) is a first-generation American-Nigerian and Native American storyteller, engaging in community-driven research. Her work examines the politics of narrative, aiming to amplify community voices and illuminate the rich tapestry of the Black experience through diverse storytelling methods.

Estefania Ciliotta Chehade (Associate Director, Center for Design, CAMD, NU) researches how facilitated interactions can improve cross-cultural team-working. She applies experience, service, and participatory design methodologies and frameworks, and her work has been used in a variety of organizations and communities to improve collaboration, trust, and teamwork.

Maria Ramirez Cifuentes (PhD student, Interdisciplinary Design & Media, CAMD, NU) has led design projects for newsrooms, renewable energy, aerospace management, and satellite deployment in Ecuador, Italy, and Spain. Her PhD research focuses on designing for participatory and transformative processes to achieve sustainable futures following natural disasters.

Ashley Colley is a university researcher at the University of Lapland, Finland, where he leads the Design study program. His research focuses on human–computer interaction (HCI), on which he has published extensively. Colley has a long industry background, including serving as a principal designer at Nokia and co-founding Oura Health.

Oka Daisuke is an environmental designer engaged in urban development, world heritage and national historic sites, parks, architecture, signage, and graphics. He worked toward the founding of Omuta City Tomodachiya Picture Book Museum and now serves as its director, bridging design expertise with cultural and community development.

Jo van Engelen, University of Technology Delft and University of Groningen, The Netherlands. Van Engelen served as Emeritus Full Professor of Product Development and Strategy in Groningen since 1991 and of Integrated Sustainable Solutions in Delft since 2011. As well he served on two executive boards and 20 non-executive boards, with a focus on innovation and technology.

Zhengya Gong is a postdoctoral researcher at the Faculty of Art and Design, the University of Lapland, Finland. Her research focuses on topics such as extended reality (XR) and human–computer interaction, as well as design creativity, creative methods, design cognition, design thinking, and the influence of culture on design.

Hanna Gröndahl (ThM) is pursuing an M.Sc. in Computer Science at Tampere University, Finland. Her research interests include network security, secure processes, and human-technology interaction.

Jonna Häkkilä is Professor for Industrial Design at the University of Lapland and a docent in Human Computer Interaction at the University of Oulu, Finland. She leads the user experience research group LUX focusing on design and technology and has received numerous grants, e.g., from the EU, Academy of Finland, and Interreg.

Henna Huotarinen is a specialist working in the Frostbit Software Lab of the Lapland University of Applied Sciences. Her research interests include visual design, UX design, and virtual productions. As a master's graduate in graphic design, Henna focuses on combining visual quality and user-friendliness in all her work.

Kuisma Hurtig, MA, is a product and service designer with the Lapland User Experience group (LUX). His work combines design with research, creating solutions that are simple, practical, and meaningful. He has a strong interest in the circular economy, and outside work can often be found running or snowboarding on local hills.

Tokushu Inamura is a designer/engineer born in Aotearoa New Zealand with roots in the Amami Islands and is an Assistant Professor at Kyushu University since 2016. Holding a Double Masters Degree in Innovation Design Engineering from the Royal College of Art and Imperial College London, he specializes in Post Human-Centered Design.

Rita Jonyer, Avinor AS, Norway. Jonyer is circular economy advisor at Avinor Oslo Airport. She works with mapping out waste streams to develop and implement measures to reach Oslo airport's zero-waste goals. She and the airport's circular team work to achieve documented results in food waste reduction, increasing source separation, reducing consumption, and increasing reuse.

Matilda Kalving is a doctoral researcher at Tampere University studying human–AI collaboration in creative tasks. With a background in design, her interdisciplinary work examines how AI can enhance creativity and technology, particularly in design, film, and writing, contributing to progress in both academic research and industry practice.

Jyrki Kaukomies is a senior expert at Content Union, specializing in futures foresight, speculative design, and service process development. With a focus on design fiction methods, he works at the intersection of design and futures-oriented business development, creating innovative approaches to engagement, participation, and strategic transformation.

Marjuska Kaukomies holds a Master of Arts degree and works as CEO and senior expert at Content Union, specializing in service design, customer insight, and interaction environments. She collaborates with academic and organizational partners to develop research-based, sustainable solutions. Her work bridges empathy, foresight, AI, and interdisciplinarity to enable bold and radical innovation.

Pigga Keskitalo is Professor of Education at the University of Lapland, specializing in Arctic education perspectives and a docent in Multicultural Education at the University of Helsinki, Finland. She leads the projects LINCOSY (Academy of Finland) and LOHKANLIHKKU (Interreg Aurora) at the University of Lapland. She also serves as a member of the Strategic Research Council (SRC) in Finland.

Chorong Kim is an Assistant Professor of Visual Studies, Kongju National University, South Korea. Her research interests include social innovation, public policy design, civic engagement, and business consulting. She worked with social enterprises and startup incubators, leading projects in impact branding, IR Strategy, and marketing content design.

Miso Kim, Assistant Professor of Experience Design at Northeastern University, studies service design through humanist frameworks, focusing on dignity, autonomy, and participation. She holds a PhD and two MDes degrees from Carnegie Mellon and a BS in Architecture from Sungkyunkwan University and serves as Design Director of the NuLawLab.

Emma Kirjavainen is a doctoral researcher in the Faculty of Art and Design at the University of Lapland. As a member of the LUX research group, she explores the intersection of human-robot interaction, user experience, and design, investigating how empathy and emerging technologies can enhance human-centered design practices.

Reetta Kupari is a seasoned expert in service design, and she is an inspiring educator at the University of Lapland, as well as in a multidisciplinary, hybrid-format master's program. She leverages visualization to ignite interaction, foster understanding, and strengthen connections. To her, every encounter is an opportunity for discovery and insight.

Hong Li is an Associate Professor of Service Innovation and Digitalization at the University of Stavanger in Norway. She holds a doctorate with distinction in Human-Computer Interaction. Her primary research explores the (re)design of intensive technologies to foster positive human experiences.

Ville Louekari (MA) is a doctoral researcher at the Centre for European Studies at University of Helsinki, Finland. He works in the fields of philosophy and European studies. His research focus is on the temporal aspects of environmental politics.

Timo Luukkonen, MA in Industrial Design and BSc in IT Engineering, is a member of the Lapland User Experience (LUX) research group. His research focuses on human–computer interaction (HCI), prototyping, and evaluation of immersive, human-centered environments.

Michael Arnold Mages (Associate Professor, Center for Design, CAMD, Northeastern University) has developed designs for civic conversations with the Cities of Boston, Pittsburgh, and US Department of Veterans Affairs. He is the author of the book *Conversational Design*. His research—funded by Heinz Endowments, Mellon Foundation, and NIH—explores facilitating sense-making in difficult situations.

Ravi Mahamuni is Principal Scientist at TCS Research and Adjunct Professor at IDC School of Design, IIT Bombay. He is a design thinker with 29 plus years of experience; his research focuses on design for change, system design, and service design at scale. He believes in technology as an enabler while humanizing the services. He also advises several academic institutions on matters related to design, technology, and social change.

Jouko Makkonen is a doctoral researcher at Tampere University. His PhD research explores residents' experiences and expectations of intelligent technologies that support space sharing and communality, particularly in Nordic Superblock type neighborhoods. He works in the Human-Centered Technology (IHTE) research group in the Computing Sciences unit.

Satu Miettinen is currently the Dean of the Faculty of Art and Design at the University of Lapland in Finland. She is also a Professor of Service Design at the same university. Satu Miettinen is recognized as one of the pioneers in the field of service design. She is leading the service design research group.

Ki-Young Nam, Associate Professor of Industrial Design at KAIST. His research interests lie in design approaches to social innovation. He has served as an editor, executive, PC, and reviewer for international/domestic academic organizations, including Vice President of Korea Society of Design Science. He is leading SIG Social Impact in KSDS.

Petri Nokelainen is a Professor of Engineering Pedagogy at Tampere University, Finland. His research interests cover teaching and learning in engineering higher education, professional growth, workplace learning, and technology-enhanced learning.

Siiri Paananen is a university teacher and doctoral researcher at the University of Lapland, User Experience Design group, working in Human–Computer Interaction. Her research focuses on augmenting cultural heritage experiences with interactive technologies and engaging the users and stakeholders through co-design methods.

Sofie Pelsmakers is Professor of Sustainable Housing Design at Tampere University where she co-chairs the ASUTUT research group. Through her internationally recognized and awarded research, and education of architects, she aims to influence the design of sustainable living environments and resilient communities now and in the future.

Kati Pitkänen is a human geographer and senior researcher at the Finnish Environment Institute. She studies how people and communities shape a sustainable future, focusing on citizens' and consumers' roles in the circular economy, social sustainability of the circular economy, and how nature supports wellbeing.

Manuela Quaresma is a researcher and professor at PUC-Rio with expertise in Human Factors, Human-Centered Design, and Service Design. Her experience spans user research, UX, and co-design across physical and digital interfaces. She leads strategic design projects, focusing on solving complex problems through collaborative, user-centered, and innovation-driven approaches.

Nur Samancioglu is a postdoctoral researcher in Human-Centered Technology research group at Tampere University. Her research focuses on evaluating and enhancing the technological readiness of built environment, integrating human-centered design, and developing technology-driven solutions for sustainable, user-friendly urban environments that bridge the gap between technology, architecture, and human experience.

Melanie Sarantou is a Professor of Social Design at Kyushu University in Japan. Her interests span the role of improvisation in arts and design practices when the unsuspected emerges.

She seeks to understand how the haptic and tactile qualities of arts and narrative practices can sustain marginalized communities.

Heloísa Seratiuk Flores is an Assistant Professor at the Kyushu University Institute for Asian and Oceanian Studies. Her research focuses on translating scientific advancements into practical healthcare applications through design. Her current projects explore the embodiment of medical devices, as well as the implementation of healthcare systems in remote areas.

Jules Rochielle Sievert, Creative Director of NuLawLab since 2013, earned a Ph.D. in Northeastern University's Interdisciplinary Doctoral Program. Their dissertation advances a transformative, transdisciplinary legal design framework. Sievert was Primary Investigator for NuLawLab's NEA Spatial Justice research and is currently Co-Principal Investigator on a Mellon Foundation–funded special initiative

Mari Suoheimo, The Oslo School of Architecture and Design, Norway. Suoheimo works as an Associate Professor, and her main research focus is on how to approach wicked problems in service design. Suoheimo is also part of the editorial board of the *Design Journal* and the *Contexts Systemic Design Journal*.

Kanako Tomisawa (Ph.D., Em.D.) is the Spokesperson and Educator of Omuta City Zoo. She contributes to IUCN/SSC/CPSG, IUCN/CEC, International Zoo Educators Association (Board member), South East Asian Zoos and Aquariums association (Species Management Committee Secretariat, Conservation Education Committee Member), RED PANDA NETWORK JAPAN (Board member), as well as an adjunct lecturer of Kyushu University.

Elisabeth Tschavgova, University of Technology Delft, The Netherlands. Tschavgova is a Ph.D. candidate at the Industrial Design Faculty at TU Delft. Her research explores systemic design methodology in a business context to develop interventions addressing complex problems related to the climate crisis. Through action-based research, she analyzes systemic design approaches.

Tuukka Uusitalo has been working in photography and filmmaking since the 1980s. He studied photography, fine arts, and cinematography at Aalto University. He´s active in the film industry as a cinematographer at Zone2 Pictures Ltd & Zone ii Ltd and a senior lecturer at Oulu University of Applied Sciences.

Kaisa Väänänen is Professor of Human-Technology Interaction in Tampere University, Finland, where she leads the research group of Human-Centered Technology. Kaisa has three decades of experience of interdisciplinary research both in industry and academia, focusing on Human-Centered AI design research and sustainable development. Kaisa is an ACM Distinguished Member.

Sonja van Dam, University of Technology Delft, The Netherlands. Van Dam is Assistant Professor in Circular Product Design and Transitions and Energy Consumption at the Faculty of Industrial Design Engineering. Key elements in her work are co-creating circular product-service systems and designing for sustainable behavior change.

Minna Virkkula is a Senior Lecturer at Oulu University of Applied Sciences and a doctoral researcher at the University of Lapland. With a strong background in UX design and research in industry, her current work explores safety design in services, user experience research, and service design development in complex environments.

YangYang Zhao is an Associate Professor of Digitalization and Entrepreneurship at the University of Oslo. Her recent research interests include fuzzy-front-end design, human–AI co-evolution, and responsible digital transformation. With experience across Europe, Asia, and the USA, and in the IT startups, her work has appeared in many peer-reviewed journals and conferences.

Peer Reviewers

The editors would like to extend their appreciation and gratitude to the following peer reviewers of the book.

Estefania Ciliotta Chehade, Northeastern University-Center for Design, United States of America
Amalia de Götzen, Aalborg University, Denmark
Jonna Häkkilä, University of Lapland, Finland
Emma Kirjavainen, University of Lapland, Finland
Sheng-Hung Lee, Massachusetts Institute of Technology, United States of America
Markus Löchtefeld, Aalborg University, Denmark
Ravi Mahamuni, TCS Research, India
Heiko Müller, University of Oldenburg, Germany
Siiri Paananen, University of Lapland, Finland
Minna Pakanen, Aarhus University, Denmark
Jules Rochielle Sievert, Northeastern University, United States of America
Heloísa Seratiuk Flores, Kyushu University, Japan
Mari Suoheimo, The Oslo School of Architecture and Design, Norway
Niina Turtola, Edinburgh Napier University, United Kingdom
Minna Virkkula, University of Lapland, Finland

1 Introduction

Mari Suoheimo, Jonna Häkkilä, and Ravi Mahamuni

Introduction

Design and technology have the capacity to engage people – citizens, communities, participants, and users – in a multitude of different ways. Fast advances in technology development and connectivity have revolutionized our capabilities to reach out to other people and participate in activities and conversations across the globe. In today's world, supporting remote work and schooling, maintaining social relationships through online media, building resilience for pandemics and other societal crises, and enabling a more sustainable or rural lifestyle, all emphasize the importance of technology-meditated participation (Karhu et al., 2021; Van Zoonen et al., 2021; Zenkteler et al., 2021). With services and applications utilizing emerging technologies such as AI, virtual reality, and smart environments, we can expect new forms of participation to arise. Engaging citizens and communities at different levels of society paves the way for democratizing and distributing power (Suoheimo et al., 2025) and builds trust between people, organizations, and policymakers representing local people and their needs (Malek et al., 2019). This book aims to gather together topical research on the theme of participation and engagement and the roles of design and technology in it.

Participatory or citizen-centered design has a long-standing basis in engaging end-users, communities, and citizens in co-designing services (Suoheimo & Lusikka, 2020), urban planning (Geekiyanage et al., 2021), policies (Bason, 2016), smart cities (Rocha et al., 2022), and healthcare (Hamzah & Wahid, 2016), to name but a few. Participatory and co-design methods engage people, citizens, children, seniors, decision-makers, and even non-human actors. Participatory design engages diverse stakeholders in the planning, execution, and analysis of a project, with the goal of fostering engagement at every stage (Cornish et al., 2023). The aim is often to foster empathy toward these different stakeholders and ensure that their needs are met (Sarantou & Miettinen, 2022) and to promote social inclusivity and responsibility (Zamenopoulos & Alexiou, 2018). The underlying idea of participatory design is that by involving people in the design process, their needs and preferences are integrated into the development of the products, services, or systems, and they are more likely to also adapt and use them. The participatory approach also expands to organizations and institutions, which can similarly take part in the process of engagement and knowledge-building activities. The designers often have roles in negotiating the needs of the different actors when creating the overall solutions.

Numerous research approaches that apply engagement with actors exist, including participatory action research, participatory design, research through design, research by design, co-design, and artistic research. In the design field, researchers and practitioners may use design thinking (Brown, 2008), systems thinking (Suoheimo et al., 2025), or the systemic design approach, and methods such as double diamond (Design Council, 2025), evolved double diamond (Design

DOI: 10.4324/9781003667872-1

Council, 2020), and triple diamond (de la Harpe & Ogundaini, 2023). All of these are used to tackle the challenges and future development directions for the society and organizations, as well as future service and product development. The complexity of the research question and the framing of the research context often reveal what approach is the most appropriate. The choice of research approach is also influenced by the research tradition the researcher group is coming from. These models help in structuring the research and design flow and in communicating the fuzzy processes of participation and engagement for the actors not familiar with the design processes.

This book contains chapters that demonstrate how technology has enabled engagement and reimagined citizen participation today and for the future. The chapters range from theoretical research to case studies, including works that take a bold approach to rethinking engagement. The participatory design approach that emerged in the Nordic countries (Gregory, 2003) has since spread to different parts of the globe and is widely used (Bødker et al., 2022), as demonstrated in this book's chapters, which explore the meaning of engagement in various cultural contexts. This book mostly covers the perspectives on participatory design, co-design, and engagement from the Nordics and Europe but also includes works from other regions such as South America, North America, and Asia. When practicing participatory design, one needs to be aware of situational knowledge arising from geographical, historical, cultural, and temporal contexts (Paananen et al., 2022; Dittrich et al., 2014).

This book addresses technology as part of the design context. The extremely wide landscape of technology makes defining and confining it to a distinct box impossible, and its ever-advancing usage has been integrated into all sectors of life, including both tangible and intangible forms. Tangible interfaces and physical objects still prevail as a gateway in accessing complex services and data structures, and mobile phones and access to the internet are omnipresent in most of our everyday lives. The development of new solutions for any sector of society can hardly be thought of without considering also technological enablers. Technology also bridges different research fields and is today an essential part of research practice, allowing efficient and large-scale data collection. When thinking of the future development trends, acknowledging the down-sides of technology use, as well as its impact on planetary aspects, is also critical. Although often technologies are designed to aid humans, they also affect non-humans.

This book contains four subsections focusing on the **landscapes of engagement and participatory design democracy, sustainable ways of living, smart cities, and energy and transport.** This book has design professionals from a broad spectrum: interaction and user experience design, product design, service and system design, social design, and design for policy and governance, among others, who have taken up both theoretical and case studies that create innovations in making and designing for citizen engagement. This book looks more closely at the notion of engagement in participatory activities and the role of technology in it. All chapters are using aspects of applying (emerging) technologies as a form of engagement and expand our understanding of participatory approaches in a changing world, where many of the current paradigms are being challenged. We hope that the chapters will offer you as a reader new and fresh insight into engagement as a way of doing participatory design.

Section I – Looking at the Landscapes of Engagement and Participatory Design

This book starts with a section providing a high-level overview of the engagement and participatory approaches and what their role is in the ever more technical society. First, a literature review on prior art is presented, and then a living lab framework of integrating law and future technology considerations in future society is introduced.

Chapter 2: Sharing Power by Participant Co-research: A Scoping Review Investigating the Interconnection of Participatory Action Research, Co-design, Engagement, and Technology

The chapter by Aejmelaeus, Louekari, Gröndahl, Seratiuk Flores, Suoheimo, and Nokelainen examines how the concepts of participatory action research, co-design, engagement, and technology are interrelated in contemporary scientific literature. This scoping review analyzed a total of 32 publications via thematic analysis and researcher triangulation. It is important to examine in more detail how participatory design and co-design are employed to engage people in technological development and to promote an understanding of technologies and their potential for civic participation. The authors found that technology-supported approaches, including the use of reflective tools and user feedback, demonstrate that actively involving users increases the acceptance and practical relevance of technology in daily life.

Chapter 3: Living Law Labs as Sites of Participatory AI: Experimental Infrastructures for Ethical AI Futures

In their chapter, Sievert, Ang, and Kim critically examine how one could actively co-produce structural inequalities in the context of law and technology. Using both in a responsible manner could promote citizen engagement. AI is also recognized as a tool for innovation and efficiency. However, people using AI need to be aware of the potential issues, as its algorithms could reproduce asymmetries of power and exclusion. Understanding the dangers, the authors draw from transdisciplinary perspectives and the use of the Living Law Lab framework to aim to mitigate the harm.

Section II – Engagement in Shaping Democracy

This section reports new ways for citizens to influence their future living environment and shape a better-functioning society. Chapters introduce experiences on engaging people with participatory budgeting, designing safe environments, and fair perspectives on green transition. The case studies illustrate how technologies related to generative AI and gamification can offer new tools for engagement and democracy.

Chapter 4: Democracy by Design: Applying Latin American Lessons to Boston's First Participatory Budgeting Process

The chapter by Ramirez, Cifuentes, Ciliotta Chehade, and Mages shows how design can shape or democratize decision-making about budget allocation, using a set of practical case studies from Boston. A total of 560 residents participated in the workshops, generating 1,238 ideas, making the collected data vast. Some of the findings highlight the importance of transparency and inclusion, as well as addressing inequalities. The chapter demonstrates in practice how power can be shared with citizens.

Chapter 5: Engaging in the Safety Design of Sustainable Tourism Services: The SafetySpinner Tool for Participatory Design

Virkkula, Li, and Gong are interested in safety issues when designing tourist experiences. This chapter's case study explores the use of a tool called SafetySpinner for facilitating visual dialogues with various stakeholders. The authors' findings demonstrate how the tool raises awareness of safety issues among participants and recognize that it is a user-centered process. By fostering emotional engagement and trust, the tool promotes sustainable practices and inclusive development within the tourism sector.

Chapter 6: Participatory Citizen Engagement in the North: Exploring Green Transition Futures with Generative AI

In this chapter, Paananen, Äärelä-Vihriälä, Keskitalo, and Häkkilä highlight how AI can provide new tools for citizen engagement, thus democratizing participation. However, AI can introduce bias or misleading interpretations if used recklessly. In this case study, the authors explore how pictures from the past, present, and future, created with generative AI, can be used to encourage discussion and engagement with the regions in which people are living in the Arctic. The chapter argues that AI provides powerful tools for amplifying voices. This work reconceptualizes participatory design by employing generative AI not merely as a production tool but also as a catalyst for futures thinking and local intergenerational dialogue. The generative AI tools used for the figures in this chapter are OpenAI DALL·E (via ChatGPT, model GPT-4o) and Adobe Firefly (Firefly Image 3).

Chapter 7: Rethinking Civic Engagement

The chapter by Kim and Nam looks at how silent citizens can be included in policymaking. They use citizen-centered design thinking and technology-mediated tools to encourage engagement. The authors present three practical case studies that have employed these approaches to reduce the barriers to citizen engagement. Interestingly, the gamified solutions presented in the chapter could be used to create interventions that promote equal participation while providing psychologically safe environments.

Section III – Holistic Sustainability and Way of Living

Sustainability aspects require increasingly much attention in the future, as the world is facing a serious environmental crisis as a consequence of human actions. Future visions must be expanded to consider both environmental and social sustainability. The following chapters present how participatory design can build more sustainable societies by addressing innovations in health care and recycling and reuse and by also involving more than human stakeholders in design projects.

Chapter 8: From Design to Adoption: A Responsible Approach to Stakeholder Participation in Digital Health Innovations

This chapter from Li and Zhao investigates how participatory design can strengthen the bottom-up initiatives to engage stakeholders from various backgrounds using technologies for health innovations. The case study was conducted in a Chinese hospital with 250 postoperative patients, examining specifically aftercare that is often overlooked. The findings show that engaging individual stakeholders helps address responsibility concerns and empowers them to influence how these technologies are implemented. The chapter provides recommendations to help guide future initiatives more effectively.

Chapter 9: Usability in Sustainable Systems: A Service Design Perspective

Hurtig, Luukkonen, Pitkänen, and Colley examine how service design can promote sustainable behaviors in everyday consumer life by adapting existing infrastructure to support new reuse practices. The chapter presents a case in which reusable shopping bags can be deposited and returned in stores. The designs were developed by engaging users in the process, incorporating technological approaches to address the challenge at hand. The results indicate that fostering sustainable behavior depends not only on policy and incentives but also on designing interactions that leverage everyday habits and tangible service elements to shape user experiences.

Chapter 10: Designing for Green Transitions: A Case Study of Reusable Takeaway Bowls in Airports

Tschavgova, van Dam, Bakker, van Engelen, and Jonyer investigate Oslo Airport's early transition toward sustainable practices through a systemic design case on reusable takeaway bowls. With regulations pushing back against the throwaway culture, the study engaged travelers and staff using multiple data collection methods to create interventions. The chapter highlights challenges such as technological scalability, limited consumer awareness, and low employee motivation. Authors notice that a systemic approach for designing integrated pricing, logistics, and user journey improvements is essential.

Chapter 11: Digital Archiving with 3D Photogrammetry and On-Site Ideation for Enhanced Participation with Non-humans

The authors, Inamura, Sarantou, Tomisawa, and Oka, examine new forms of engagement involving guinea pigs, humans, artificial intelligence (AI), and technology in a zoo in the process of participatory development. Through post-humanist arts-based research, the study examined how different stakeholders could work together to design and develop more sustainable enclosures, encouraging interaction with animals in zoological gardens through participatory activities. The findings suggest that combining low-fidelity prototypes with rapid digital photogrammetry can generate ideas and solutions to ethical care challenges in zoological settings.

Section IV – Future Smart Cities and Transportation

The chapters in this section imagine future urban environments and services and address what these futures could look like and how to design for them by engaging different stakeholders. Perspectives into designing mobility, services, and integration of smart technology in everyday smart city environments are presented.

Chapter 12: Designing Mobility Futures through Strategic Citizen Engagement

The chapter from Quaresma looks at the transportation and mobility solutions in Brazil, Rio de Janeiro. The case study in question investigates how to engage citizens in making future visions of mobility for the years 2030 and 2050. Backcasting has been applied as a tool to understand how these visions could be reached. The study in question shows how combining both futures and co-design can serve to democratize mobility planning in complex transportation contexts.

Chapter 13: Engaging Citizens in Robot Dog Interaction in Smart Cities – Perceptions in Public Spaces

Kirjavainen, Kalving, Paananen, Colley, and Häkkilä explore the possibilities of using social robots as robot dogs in urban settings. The data consists of 100 survey responses and 21 field interviews with citizens investigating the acceptance of using robot dogs in environments such as shopping malls, cafes, or museums. The research participants thought that roles such as security, navigation, and cleaning would be suitable. The responses also show how the context plays a crucial role in whether a robot can be accepted to have interaction with citizens. This study advances the understanding of engagement by showing how participatory, emotional, and situational factors converge in human-robot interaction.

Chapter 14: Engaging Stakeholders in Prototyping with Immersive Theatre Practices and Worldbuilding on the Virtual Production Stage

The chapter by Björn, Miettinen, Uusitalo, Kaukomies, Kaukomies, Huotarinen, and Kupari investigates how companies could use, to a larger degree, virtual production stage as a possibility for prototyping near-future services. Using performing as a form of prototyping is one way to create speculations of embodied user engagement. Emotions and experiences play a crucial role when bridging the imagination and implementation. This chapter advances the discussion on using empathy to enrich immersive and narrative approaches for imagining and co-creating emotionally resonant futures.

Chapter 15: A Framework of Engaging Smart Technologies for Citizen Participation in the Lifecycle of Sustainable Built Urban Environments

Väänänen, Samancioglu, Makkonen, Castano De la Rosa, Autero, and Pelsmakers couple cities' resources, energy, and social equity challenges when designing innovative solutions for urban built environments. Interactive digital tools are essential to move citizens beyond serving only as informants during the planning stage, enabling them instead to become active and sustained contributors to the development of urban spaces. This chapter introduces UrbanEngagingTech, a framework that enables citizen participation from buildings to cities across all stages of the urban lifecycle.

The chapters in this book have a vast variety of topics. We hope you can find chapters that are related to your research or development interests.

We wish you good reading time.

The editorial team,
Mari Suoheimo, Jonna Häkkilä, and Ravi Mahamuni

References

Bason, C. (2016). *Design for Policy*. Routledge.

Bødker, S., Dindler, C., Iversen, O. S., & Smith, R. C. (2022). What can we learn from the history of participatory design? In S. Bødker, C. Dindler, O. S. Iversen, & R. C. Smith (Eds.) *Participatory Design* (pp. 15–29). Cham: Springer International Publishing.

Brown, T. (2008). Design thinking. *Harvard Business Review*, *86*(6), 84.

Cornish, F., Breton, N., Moreno-Tabarez, U., Delgado, J., Rua, M., de-Graft Aikins, A., & Hodgetts, D. (2023). Participatory action research. *Nature Reviews Methods Primers, 3*(1), 34.

de la Harpe, R., & Ogundaini, O. (2023). A proposed transformation service design research framework for underserved settings. In *Service Design for Emerging Technologies Product Development: Bridging the Interdisciplinary Knowledge Gap* (pp. 73–86). Cham: Springer International Publishing.

Design Council. (2020). *What Is the Framework for Innovation? Design Council's Evolved Double Diamond.* Available from: www.designcouncil.org.uk/news-opinion/design-process-what-double-diamond (accessed on 8 Sep 2021).

Design Council. (2025). *Beyond Net Zero—A Systemic Design Approach.* Available from: www.designcouncil.org.uk/fileadmin/uploads/dc/Documents/Beyond%2520Net%2520Zero%2520-%2520A%2520Systemic%2520Design%2520Approach.pdf (accessed on 25 Mar 2025).

Design Council. (2025). *From Humble Beginnings to a Cornerstone of Design Language.* Available from: www.designcouncil.org.uk/our-resources/the-double-diamond/ (accessed on 26 Sep 2025).

Dittrich, Y., Eriksén, S., & Wessels, B. (2014). Learning through situated innovation. Why the specific is crucial for participatory design research. *Scandinavian Journal of Information Systems, 26*(1), 2.

Geekiyanage, D., Fernando, T., & Keraminiyage, K. (2021). Mapping participatory methods in the urban development process: A systematic review and case-based evidence analysis. *Sustainability, 13*(16), 8992.

Gregory, J. (2003). Scandinavian approaches to participatory design. *International Journal of Engineering Education, 19*(1), 62–74.

Hamzah, A., & Wahid, F. (2016). Participatory design in the development of healthcare systems: A literature review. In *Proceedings of the 2nd International Conference on Communication and Information Processing* (pp. 60–64). New York: ACM.

Karhu, M., Suoheimo, M., & Häkkilä, J. (2021). People's perspectives on social media use during the COVID-19 pandemic. In *Proceedings of the 20th International Conference on Mobile and Ubiquitous Multimedia* (pp. 123–130). New York: ACM.

Malek, J. A., Lim, S., & Tahir, Z. (2019). Understanding the issues of citizen participation. *Journal of Nusantara Studies (JONUS), 4*(1), 1–22.

Paananen, S., Suoheimo, M., & Häkkilä, J. (2022). Decolonizing design with technology in cultural heritage contexts-systematic literature review. In *Congress of the International Association of Societies of Design Research* (pp. 1839–1855). Singapore: Springer.

Rocha, N. P., Bastardo, R., & Pavão, J. (2022). Citizens' participation in the co-design of smart cities' applications, a scoping review. In *Proceedings of the 10th International Conference on Software Development and Technologies for Enhancing Accessibility and Fighting Info-exclusion.* (pp. 172–177). New York: ACM.

Sarantou, M., & Miettinen, S. (Eds.). (2022). *Empathy and Business Transformation.* London: Taylor & Francis.

Suoheimo, M., & Lusikka, T. (2020, August). Process for mapping challenges of cross-border mobility in the Barents region. In *International Conference on Design Creativity* (pp. 168–177). Oulu, Finland: Design society. https://doi.org/10.35199/ICDC.2020.21

Suoheimo, M., Jones, P., Lee, S. H., & Sevaldson, B. (2025). *Systemic Service Design* (p. 308). Taylor & Francis.

Van Zoonen, W., Sivunen, A., Blomqvist, K., Olsson, T., Ropponen, A., Henttonen, K., & Vartiainen, M. (2021). Factors influencing adjustment to remote work: Employees' initial responses to the COVID-19 pandemic. *International Journal of Environmental Research and Public Health, 18*(13), 6966.

Zamenopoulos, T., & Alexiou, K. (2018). *Co-design as Collaborative Research.* University of Bristol/AHRC Connected Communities Programme.

Zenkteler, M., Foth, M., & Hearn, G. (2021). The role of residential suburbs in the knowledge economy: Insights from a design charrette into nomadic and remote work practices. *Journal of Urban Design, 26*(4), 422–440.

Section I

Looking at the Landscapes of Engagement and Participatory Design

2 Sharing Power by Participant Co-Research

A Scoping Review Investigating the Interconnection of Participatory Action Research, Co-design, Engagement, and Technology

Monica Aejmelaeus, Ville Louekari, Hanna Gröndahl, Heloísa Seratiuk Flores, Mari Suoheimo, and Petri Nokelainen

Introduction

The widespread use of digital devices and applications such as smartwatches, phones, computers, and televisions offers new avenues for enabling participation and engagement. Participatory research is well established in design by engaging people in various processes, particularly in contexts such as healthcare application development. Digital technologies now play a central role in enabling engagement, with video conferencing platforms, for instance, supporting continuity between participatory action research (PAR) cycles. Both PAR and co-design are increasingly employed in technology-oriented studies to enhance user engagement. However, limited research has examined the combined use of PAR, co-design, and engagement within technology-related contexts.

This scoping review maps existing literature to examine the intersection of PAR, co-design, and engagement in technology-related contexts. The aim was to understand how these terms "participatory action research", "co-design", "engagement", and "technology" are used together when conducting academic research. In this manner, our research question is:

> How have participatory action research, co-design, and engagement been applied in technology-related contexts?

Following the PRISMA-ScR guidelines, a systematic search of academic databases was conducted in March 2025, identifying peer-reviewed journal articles, conference papers, and book chapters. Studies were included if they employed both PAR and co-design while addressing technology and engagement. Thirty-two studies were identified across the fields of healthcare, research methodology, human-computer interaction, communication, and sustainability. The limited number of studies identified suggests that the combined use of these four terms remains limited in academia, indicating a research gap and potential for novel knowledge creation.

This chapter begins with looking at the recent literature and research related to co-design with technologies and similarly PAR, engagement, and technologies. In the methods section, we will detail how the scoping review was conducted. We will present the findings using charts and figures and examine thematic overlaps identified during the research. Our study shows how research on the integration of PAR in co-design studies focused on technology and engagement is still limited. Closely related work includes Benz et al. (2024), who documented the full application of co-design in a community-based participatory research (CBPR) telepractice

DOI: 10.4324/9781003667872-3

project, highlighting the implications of design choices. Gingell et al. (2024) illustrated co-design through the collaborative development of data collection tools with refugee community members in a PAR context. Similarly, Tariq et al. (2023) employed a virtual CBPR approach to engage co-researchers and examined associated methodological challenges. By investigating the intersection of PAR, co-design, and engagement in technology-related contexts, this review offers novel insights into methodological integration and user-centered technology development. Following Webster and Watson's (2002) guidance, the review seeks to advance theory in this emerging area.

Co-Design with Technologies

Co-design is a collaborative design approach that actively involves stakeholders in developing solutions (Sanders & Stappers, 2008; Vargas et al., 2022), including those directly affected by the outcomes (Trischler et al., 2019). Initiatives may originate within or outside the affected community. The approach aims to produce relevant, improved, and meaningful solutions (Trischler et al., 2019) while promoting social inclusivity and responsibility (Zamenopoulos & Alexiou, 2018). It also facilitates the identification of users' unarticulated or latent needs (Trischler et al., 2019).

Stakeholder engagement is integral to co-design and serves as a source of innovation (Zamenopoulos & Alexiou, 2018). As a knowledge-building approach, it contrasts with traditional research paradigms where researchers act as outside spectators (Steen, 2013). Engagement in co-design supports shared problem understanding, feedback collection, information gathering and analysis, and the shaping of the design process (Zamenopoulos & Alexiou, 2018). It typically begins in the early research phases (Trischler et al., 2019) and may extend throughout the process into implementation stages (Vargas et al., 2022).

Co-design is increasingly advocated in technology development projects (Treasure-Jones & Joynes, 2017). This approach has shown particular promise in the development of electronic healthcare tools, where involving end-users, such as older adults, can be fostered and enriched through the use of technology. Cole et al. (2022) suggest that higher levels of participation, potentially facilitated by various digital involvement processes, may increase users' knowledge of technology, enhance learning, and empower participants.

PAR and Technology

PAR is a qualitative research methodology (Macdonald, 2012), which focuses on social change (Fals-Borda, 2001), seeking to democratize knowledge production by grounding it in real community needs and learning (Chevalier & Buckles, 2013). PAR is characterized by a collective commitment to critically examine specific issues through self- and collaborative reflection, with the dual aim of deepening understanding and producing practical solutions for those directly affected (McIntyre, 2008). Participants are active contributors throughout the research process, not subjects of inquiry (Chandler & Torbert, 2003; Kelly, 2005). PAR's democratic ethos promotes inclusive participation and affirms the equal value of all participants (Stringer, 1996).

Collaboration, participation, and reflection are central to PAR and occur through iterative cycles of planning, action, and review (Stringer, 1996). PAR lacks a standardized design or universal theoretical framework; instead, each project is shaped by the participants' interests and aspirations. Participants choose to act on topics relevant to their context, resulting in diverse outcomes from influencing public policy and initiating informal community improvements to raising awareness of local issues (McIntyre, 2008).

In youth engagement, technology, often defined as 'youth media' and encompassing communication tools such as the internet, photography, video, and music software, serves as an effective strategy. Flicker et al. (2008) describe the e-PAR model, emphasizing how these tools attract young people by enabling them to communicate their thoughts and feelings in ways they feel will be heard and to a broader community. Gibbs et al. (2020) further note that technology expands the reach and impact of youth-led PAR, democratizing young people's access to information, research skill development, and civic participation. Digital platforms support flexible data collection and sharing, with the potential to enrich data through creative methods.

Engaging with Technologies

Technology plays a significant and multifaceted role in fostering engagement within PAR (Reason & Bradbury, 2001) and co-design (Sanders & Stappers, 2008) processes. In the context of PAR, the incorporation of digital tools and platforms can enhance dialogue, build understanding, and motivate positive actions among participants.

Paluch et al. (2023) illustrate engagement processes during the COVID-19 pandemic, where digitally mediated PAR became crucial for maintaining engagement and social connections for older adults. Enabling their participation through technology, alongside care practices and attention to their digital literacy, was essential for fostering self-directed learning and addressing their needs. Furthermore, Cockerham (2024) notes that technology can act as a leveling force in research collaborations. Technology training can help equalize perceived power hierarchies, and the digital expertise of youth co-researchers can build their confidence and place them in leadership roles. This technologically supported, non-hierarchical approach fosters open communication among stakeholders.

Despite its numerous benefits, technology also presents challenges for engagement. Gibbs et al. (2020) report difficulties in sustaining rich dialogue and collaborative decision-making during critical stages such as analysis and social action when relying solely on digital tools. They stress the importance of training adult facilitators in online engagement and addressing disparities in digital access and literacy. Similarly, Paluch et al. (2023) underscore the need to consider participants' prior technological experiences and integrate care practices to ensure meaningful engagement.

Ultimately, technology in PAR and co-design serves as a powerful tool to enhance engagement by enabling new forms of participation, communication, and learning. Its effectiveness depends on thoughtful implementation, skilled facilitation, and attention to equity and participant needs. As Tverskoi et al. (2022) note, the broader social context and attitudes toward technology significantly influence its capacity to engage participants in participatory research.

Methods

Scoping Review

Scoping reviews aim to systematically identify and map existing literature across a particular area of inquiry. Scoping reviews offer a clear indication of the volume of available studies and provide either a broad or detailed overview of the focus of the literature. Scoping reviews also serve as valuable tools for examining how research on a given topic has been designed and conducted, offering insights into methodological trends and gaps in the literature (Munn et al., 2018).

Following PRISMA-ScR guidelines, this scoping review explores how participatory action research, co-design, and engagement have been applied in technology-related contexts. The objective is to identify publications that integrate PAR, co-design, technology, and engagement.

Search Strategy

The search was conducted via Andor, Tampere University's portal encompassing 352 databases such as Scopus, Web of Science, and ERIC, in March 2025. The goal was to identify peer-reviewed journal articles, conference papers, and book chapters in English. The selected search terms were "participatory action research" AND "co-design" AND "engag*" AND "technology". The term *technolog* was initially used with a wildcard but later refined (as *technology*) to increase relevant results. A second search was conducted using codesign (without the hyphen) to ensure comprehensive coverage. The hyphenated search yielded 230 results, with 22 excluded as books, theses, or duplicates. The non-hyphenated search produced 35 results, with 12 similarly excluded.

Study Selection and Data Extraction

In total, 231 studies were screened. Google Spreadsheets was used to assess whether all four search terms were addressed in each publication. Based on this screening, 40 studies were selected for full-text review (Figure 2.1). Relevant excerpts and images were extracted into a new spreadsheet, with separate classifications for PAR, co-design, engagement, and technology.

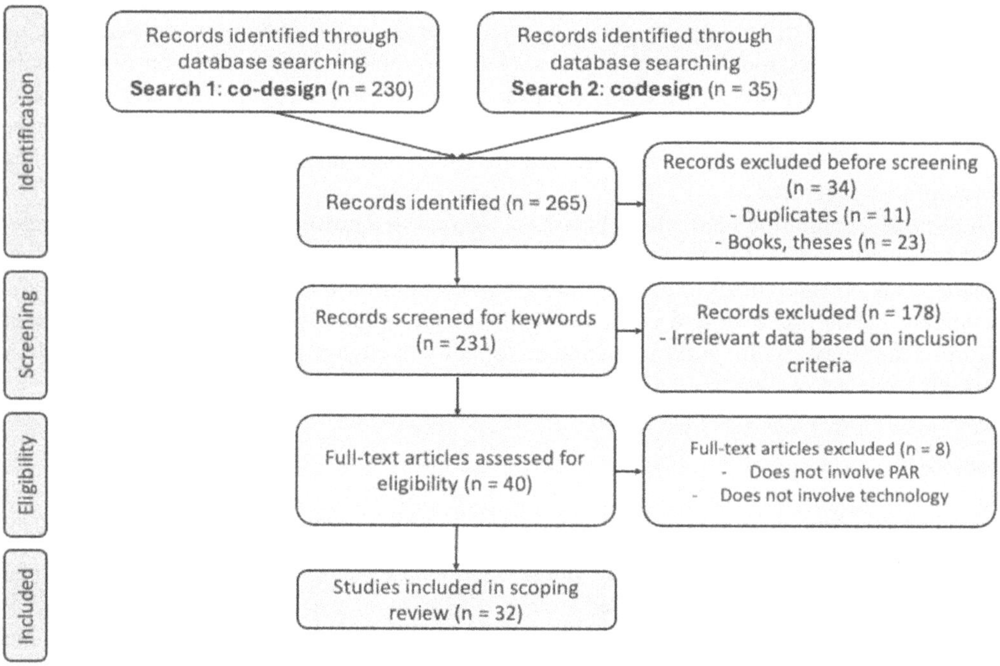

Figure 2.1 PRISMA flow diagram depicting the study selection process across the identification, screening, eligibility, and inclusion phases.

Additional data included publication year and type, study context, and participant numbers. Three researchers conducted the data extraction. Following this stage, eight studies were excluded for lack of relevance, resulting in a final set of 32 studies.

Data Analysis

A qualitative, inductive thematic analysis was conducted following Braun's and Clarke (2006) six-phase framework, detailed in the research protocol (Appendix 1). Six researchers contributed to the analysis, ensuring researcher triangulation (Carter et al., 2014). Following data extraction, four researchers conducted initial coding, after which a codebook was collaboratively developed through a peer-review-style meeting. In a three-phase process, Researchers A and B independently coded the data. Discrepancies were resolved by Researcher C, who was blinded to their responses. This multi-stage, collaborative approach helped account for differing interpretations of the data (Saldaña, 2013), acknowledging that researchers' backgrounds can influence outcomes (Long & Godfrey, 2004).

Findings

Study Characteristics

A list of included and excluded records is provided in Appendix 2. The studies included in the review were published between 2016 and 2025. Of these, eight were published between 2016 and 2019, 15 between 2021 and 2023, and 9 between 2024 and 2025. Most studies were conducted in Oceania (n = 10), followed by Europe (n = 8), Africa (n = 4), North America (n = 4), multiple continents (n = 4) and Asia (n = 2). As shown in Figure 2.2, the majority were related to healthcare (n = 18), with others addressing methodology (n = 5), human-computer interaction (HCI) (n = 4), sustainability (n = 3), and communication (n = 2).

All included studies employed qualitative methods. Seven were theoretical, including four scoping reviews, one conceptual paper, one review article, and one viewpoint article. The

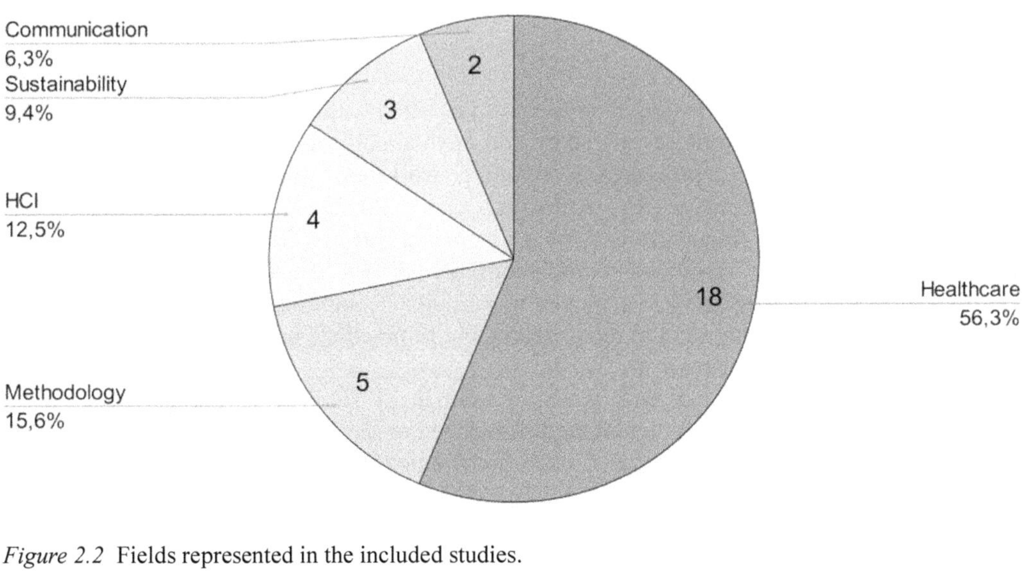

Figure 2.2 Fields represented in the included studies.

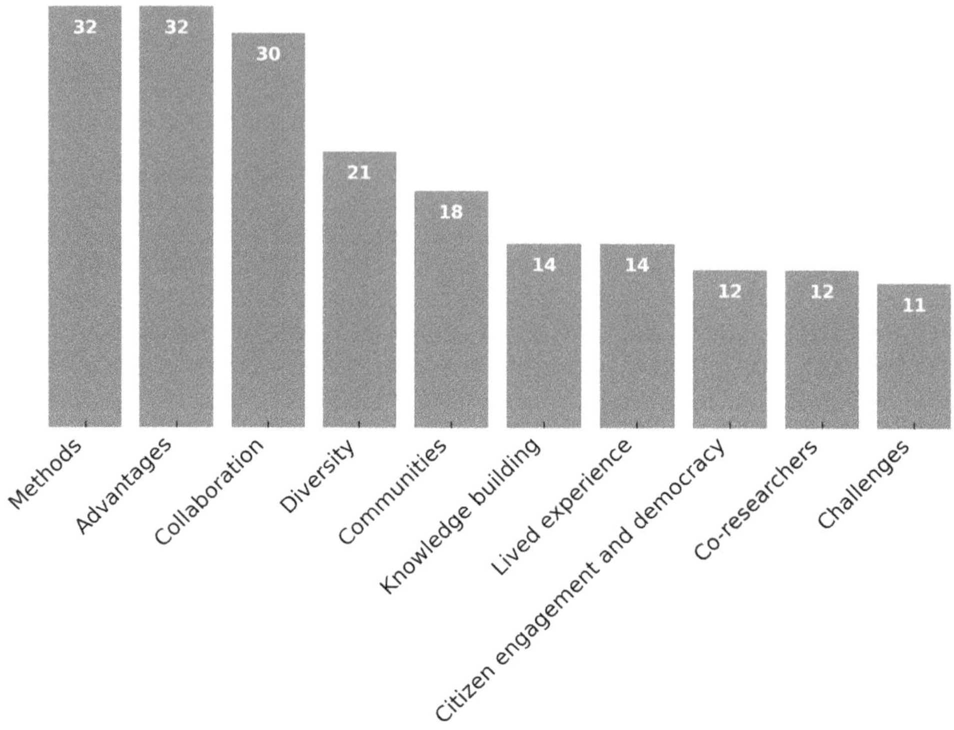

Figure 2.3 The themes identified within the context of PAR.

remaining 25 were empirical studies. Considering that some of the studies did not report the exact number of participants, the estimated number of participants engaged across the studies is approximately 5659. The four scoping reviews collectively included 419 studies.

Participatory Action Research

As shown in Figure 2.3, ten distinct themes were identified within the context of PAR. All 32 studies included in the scoping review addressed methodological aspects of PAR, employing approaches such as iterative processes, storytelling, workshops, pilot testing, and structured cycles of action and reflection (e.g., Armstrong et al., 2022; Cormick et al., 2024). The research processes were frequently described as open-ended and flexible, allowing for real-time adaptations based on continuous participant feedback (e.g., Taba et al., 2025). All studies highlighted the advantages of PAR, including participant empowerment, the promotion of equality in the research process, and the development of practical, needs-based solutions (e.g., Armstrong et al., 2022; Molapo et al., 2016).

Challenges in implementing PAR were reported in 11 studies, including difficulties in achieving genuine collaboration, managing mismatched or unrealistic participant expectations (e.g., Onwezen et al., 2021; Perry et al., 2022), and the time-consuming nature of the approach (e.g., Onwezen et al., 2021; Wallis et al., 2017). Additional barriers included high implementation costs, complexities in power-sharing, limited funding, coordination demands, recruitment challenges, and challenges in aligning the PAR outcomes to national policy-making.

Collaboration was addressed in 30 studies and community engagement in 18. The studies addressing collaboration framed PAR as a guiding framework for co-design, emphasizing practices such as collaborative analysis (e.g., Armstrong et al., 2022; Taba et al., 2025), solidarity, reciprocity, collective reflection, review, and development of tools. The studies that addressed communities included working directly with communities (e.g., Bray et al., 2022; Molapo et al., 2016), bridging academic and community practices (e.g., Onwezen et al., 2021), forming a deeper understanding of communities, and drawing on the knowledge and experiences of the affected communities.

Diversity was discussed in 21 studies, encompassing themes of inclusivity and cultural sensitivity. These studies emphasized engaging participants across age groups (e.g., Perry et al., 2022), ensuring diverse representation, and adopting culturally appropriate approaches (e.g., Perry et al., 2022; Hashim et al., 2024). Additional focus was placed on addressing the needs of minority groups, promoting decolonizing practices (e.g., Bourgeois et al., 2024; Cormick et al., 2024), fostering cultural safety, and integrating Indigenous knowledge and traditions.

Knowledge building was addressed in 14 studies, often through co-learning, active social learning, and the co-construction of knowledge within shared community projects. Rather than viewing knowledge as something to be delivered, these studies highlighted a shift toward collaborative discovery (e.g., Wallis et al., 2017). Similarly, lived experience was highlighted in 14 studies as a valuable source of experiential knowledge, underscoring the value of including those directly affected as both knowledge contributors and guides in the research process (e.g., Redublo et al., 2024).

Citizen engagement and democracy were addressed in 12 studies, as was the involvement of participants as co-researchers. Studies on citizen engagement and democracy explored themes such as developing tools to enhance citizen participation, incorporating citizen perspectives into planning, and fostering deliberative democracy. Emphasis was also placed on democratizing the research process (e.g., Devan et al., 2021; Bourgeois et al., 2024), integrating activism, and promoting social justice (e.g., Moghimi et al., 2025; Devan et al., 2021). Participant roles as co-researchers varied, including active co-creation, workshop facilitation, and contributions as citizen-scientists (e.g., Cech & Wagner, 2019). Participants were also involved in evaluating solutions and designing monitoring and evaluation processes.

Co-design

For co-design (Figure 2.4), seven distinct themes were identified. All 32 studies discussed research methods, including user engagement, collaboration, iterative cycles, workshops, toolkits, and reflection (e.g., Clark et al., 2022; Galabo & Cruickshank, 2022). Co-design was noted as particularly effective for addressing complex design problems (van Leersum et al., 2024). Advantages were identified in 31 studies, such as improved understanding of user needs, reduced bias, emancipatory impacts on communities, and the integration of lived experience (e.g., Hashim et al., 2024; Zhu et al., 2019). Challenges were reported in 23 studies, including difficulties in organizing workshops, time constraints, accommodating diverse user needs, and the absence of methodological consensus (e.g., Devan et al., 2021; Chapman et al., 2025).

Collaboration was discussed in 31 studies as a central component of co-design methodology, often involving end-user engagement to better understand community needs. The scale of collaboration varied from continuous involvement throughout the research process (e.g., Clark et al., 2022) to more limited participation in workshops, consultations, or stakeholder meetings (e.g., Hashim et al., 2024). Co-researchers were featured in 13 studies, where involving stakeholders

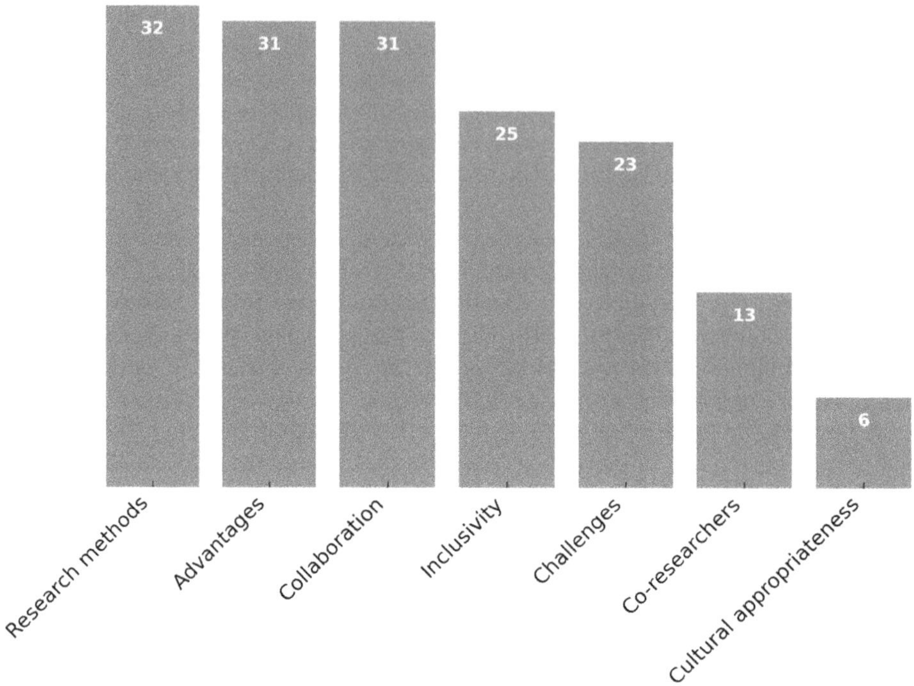

Figure 2.4 The themes identified within the context of co-design.

in the research process helped ensure the relevance of outcomes to target groups and reinforced participant agency (e.g., Clark et al., 2022).

Inclusivity was discussed in 25 studies and cultural appropriateness in six. Studies highlighted the importance of engaging directly with end-users, including individuals with disabilities, diverse age groups, and minority communities (e.g., Zhu et al., 2019). Digitalization was noted as a facilitator of access and inclusion (e.g., Perry et al., 2022). Cultural appropriateness was emphasized as essential for conducting informed, impactful research and for reducing barriers between communities (e.g., Bourgeois et al., 2024; Cormick et al., 2024).

Engagement

As Figure 2.5 shows, eight distinct themes were identified in the context of engagement. Engagement methods were discussed in 31 studies, including two-way communication, co-designed probe kits, researcher-participant dialogue, engaging in using technologies and reflexive thinking, prototype feedback, multimedia tools, peer learning, and iterative engagements (e.g., Bytautas et al., 2024; van Leersum et al., 2024). Strategies to maintain engagement between sessions or workshops were also addressed. Nineteen studies addressed engagement in research more broadly, involving participants as co-researchers in inquiries grounded in lived experience (e.g., Bray et al., 2022; Clark et al., 2022). Citizen engagement was shown to bridge the gap between science and the citizens, with active participation influencing attitudes and behavior (e.g., Taba et al., 2025) and enhancing data richness through greater openness (e.g., Redublo et al., 2024). This quotation from Bytautas et al. (2024, p. 4) illustrates how traditional research

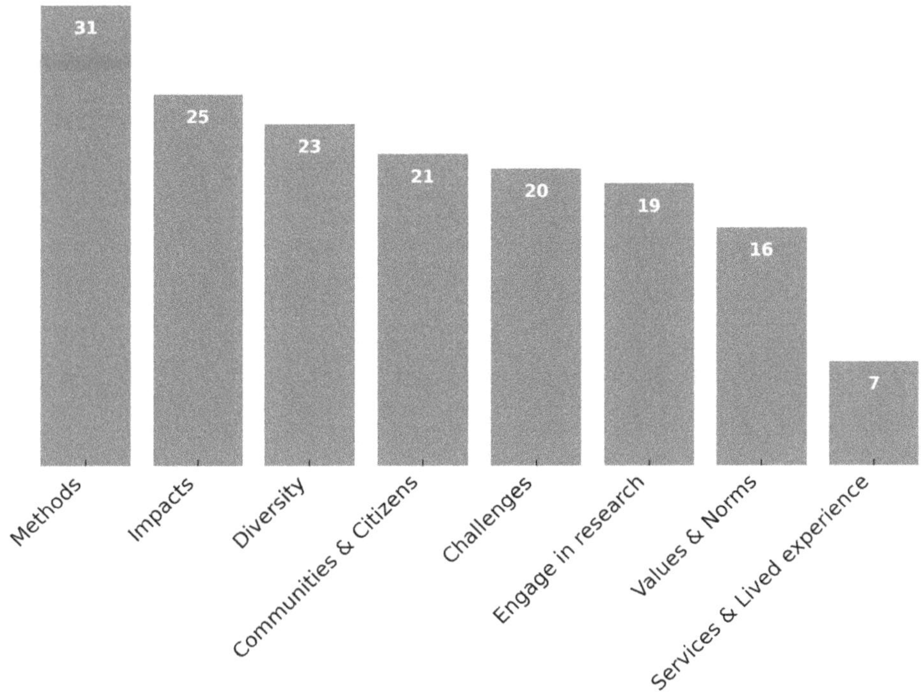

Figure 2.5 The themes identified within the context of engagement.

paradigms often overlook meaningful engagement, underscoring the need for participatory approaches based on mutual dialogue rather than hierarchical imposition:

> Kieran (AR 49): As an academic researcher myself, I do often wonder if we are too arrogant in presuming that we know what is best for people. Whether it's health or mental health, aging, social problems, or any other kind of applied field. How many researchers actually bother to speak to the people they are trying to help before moving in to solve their problems? How can we presume to know what kind of help – technological or otherwise – would be most appreciated, if we don't even have a proper conversation first?

Impacts of engagement in PAR and co-design were discussed in 25 studies, including increased participant confidence, empowerment, appreciation for being heard, adoption of preventive behaviors, emancipatory change, improved solution acceptability, equitable participation, and engagement with digital tools supporting sustainability (e.g., Bray et al., 2022; Molapo et al., 2016). Challenges were noted in 20 studies and included inconsistent engagement, difficulties engaging with the tools, design and usability concerns, and issues related to appropriate participant compensation (e.g., Chapman et al., 2025; Cormick et al., 2024).

Diversity was addressed in 23 studies, focusing on demographic representation, inclusion of people with disabilities from diverse cultural backgrounds, disability identities and experiences, engaging older adults in activities and decision-making, and enhancing engagement of marginalized groups (e.g., Bytautas et al., 2024; Chapman et al., 2025). Community and citizen

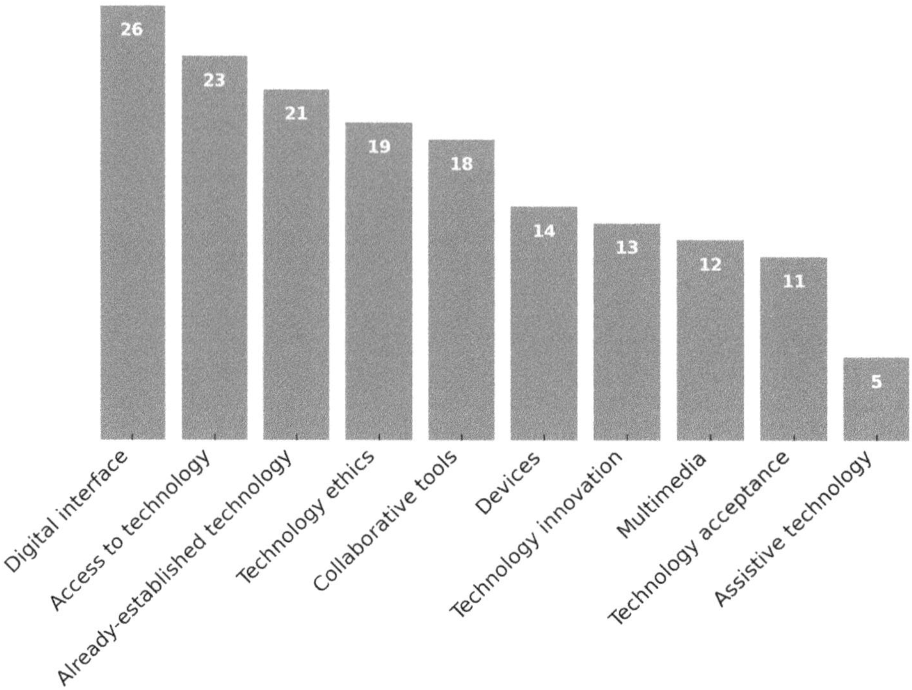

Figure 2.6 The themes identified within the context of technology.

engagement appeared in 21 studies, involving practices such as ongoing community involvement throughout research and design processes and active engagement of public participants (e.g., Cortesi et al., 2022).

Values and norms were discussed in 16 studies, including building and maintaining respectful relationships, emphasizing the importance of genuine collaboration, motivations to contribute to change, intercultural co-construction of meaning, social justice, and culturally appropriate engagement practices (e.g., Armstrong et al., 2022; Cormick et al., 2024). Services and lived experience were addressed in seven studies, focusing on patient engagement, phenomenological approaches, and the active involvement of those directly affected to ensure solutions meet real needs (e.g., Redublo et al., 2024).

Technology

Figure 2.6 shows the 11 themes identified regarding technology. Digital interfaces were discussed in 26 studies, referring to user interaction with digital systems such as apps, software, online platforms, websites, and user interfaces. Several studies highlighted tools like video conferencing platforms and online whiteboards for data collection and fostering engagement (e.g., Moghimi et al., 2025; Armstrong et al., 2022; Perry et al., 2022; Redublo et al., 2024; Bray et al., 2022). As 18 studies focused on healthcare, many digital interfaces were developed for that context, including online programs for mental health, pain management, interventions (Bray et al., 2022), and health communication tools (e.g. Hashim et al., 2024).

Access-related barriers to technology were discussed in 23 studies, where existing technologies could not be utilized due to external constraints such as lack of privacy, unstable internet connections, limited access to devices (e.g., Moghimi et al., 2025), or language barriers. Technology acceptance was addressed in 11 studies, distinguishing cases where users had access but chose not to engage, often due to exclusion from the development process, resulting in tools that did not reflect their values or needs. Negative attitudes toward technology were also linked to low digital literacy and limited opportunities for skill development (Hayes et al., 2021). Ethical concerns were raised in 19 studies, focusing on cultural implications, mental health risks, and access inequalities tied to social or economic factors. This quotation from Zhu et al. (2019, pp. 106–107) emphasizes that failing to consider the unique needs and perspectives of users can result in poor uptake and adoption of technologies:

> In addition, prior research shows that the relationship between the user, technology, and the environment should be considered in designing technology for marginalised groups [30]. Gabriels & Hill [15] suggest that applications designed for people diagnosed with ASD should allow the user to operate the device independently and that the attitude of all stakeholders involved plays a role in the application uptake success. Applications built for people with ASD should no longer focus solely on the delivery of the technology. Instead, the design approach should be inclusive of people with ASD and partner with users and communities to increase the acceptance and adoption of newly developed technology [32].

Already-established technologies were discussed in 21 studies, referring to existing tools that were refined during or after the PAR and co-design process. For example, Cormick et al. (2024) evaluated and enhanced patient journey mapping tools. Other cases involved improvements to educational technologies and materials. In contrast, 13 studies focused on technology innovation involving the development of entirely new tools (e.g., Zhu et al., 2019; Cech & Wagner, 2019).

Technologies involving or promoting collaboration were discussed in 18 studies, including creative engagement tools (Galabo & Cruickshank, 2022), educational resources for intercultural communication (Armstrong et al., 2022), and a self-organized childcare app (Cortesi et al., 2022). Multimedia technologies were featured in 12 studies, utilizing images, video, and sound. Examples included a mobile platform for community health education (Molapo et al., 2016) and online preventive healthcare programs (e.g., Hashim et al., 2024).

Assistive technology was discussed in five studies, referring to devices or systems that support individuals in performing daily functions otherwise difficult to manage independently. Device-dependent technologies were discussed in 14 studies, including mobile phones, applications (e.g., Molapo et al., 2016), wearables (van Leersum et al., 2024; Onwezen et al., 2021), and GPS trackers (Cech & Wagner, 2019). Technologies in this case were limited to gadgets.

Discussion

Employing PAR as a Framework for Co-design

The findings indicate that PAR typically serves as a guiding framework within which co-design is conducted. Research processes were often described as open-ended and adaptable, allowing researchers to respond dynamically to ongoing participant feedback, which is characteristic of PAR methodologies. As Stringer (1996) notes, collaboration, participation, and reflection are essential elements of PAR, occurring through iterative cycles of planning, action, and review.

Co-design in the reviewed studies was closely aligned with these cyclical processes. According to McIntyre (2008), PAR involves a collective commitment to critically examine specific issues through self- and collaborative reflection, with the aim of generating practical solutions for those directly affected. Participants are not passive subjects but active contributors throughout the research process (Chandler & Torbert, 2003; Kelly, 2005). The findings of the present study strongly reflected these core principles. Both self-reflection and collaborative reflection were prominent methodological components, used to support deeper understanding of the studied issue. There was also a clear emphasis on developing practical solutions that addressed the real needs of those affected. Furthermore, in many studies participants were actively involved as co-researchers, engaging in various research phases.

As shown in Figure 2.7, several themes were shared across both PAR and co-design, though the frequency of their occurrence varied. Some themes were common to both, while others appeared exclusively in one context, reflecting differences in conceptual focus and methodological orientation across the studies. For example, the shared use of co-researchers illustrates areas of overlap. Notably, challenges were more frequently reported in co-design contexts, likely due to both a higher incidence of practical issues and the tendency for publications to provide more detailed accounts of co-design practices. In contrast, PAR was often described at a more conceptual level, serving as an overarching framework.

The findings highlight the practical complexity of implementing PAR and co-design, both of which demand significant time, financial resources, careful planning, and effective communication. Although none of the included studies explicitly reported failure, many discussed challenges and lessons learned. The absence of openly acknowledged failures raises concerns

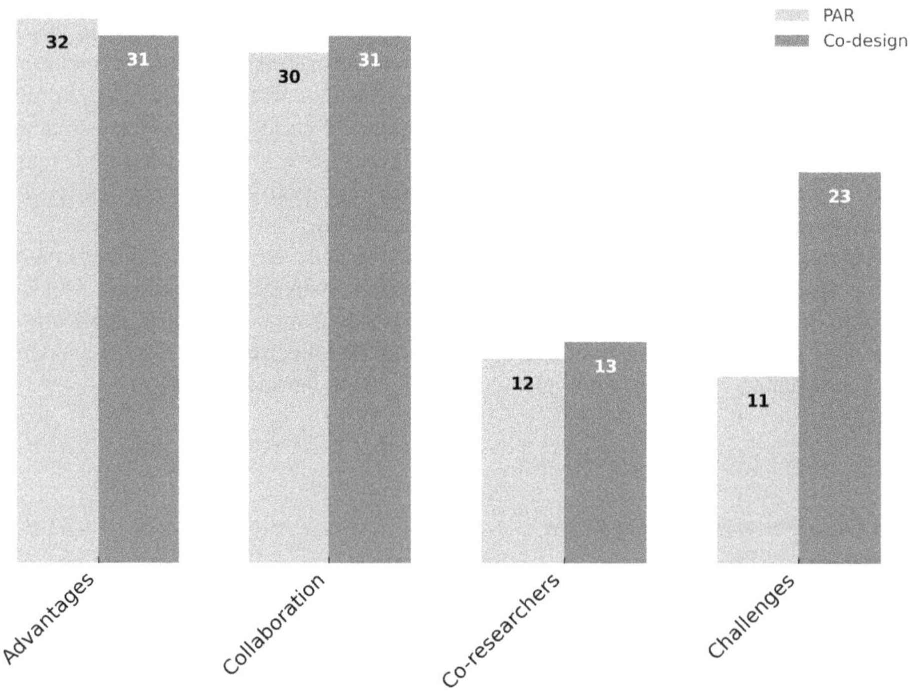

Figure 2.7 Comparison of overlapping themes in PAR and co-design.

about potential bias in reporting. While some studies demonstrated relative transparency in describing obstacles and mitigation strategies, greater openness is needed to critically reflect on and communicate such experiences in future research.

When working with participation, co-design, and the engagement of participants, particularly those in vulnerable contexts, it is important to be aware of the potential systemic oppression that may underlie a service or product, as one could be enforcing those oppressive structures (van Amstel et al., 2025). In a guest lecture, van Amstel (2024) critically reflects on the limitations of participatory and open approaches in design, arguing that empowering users or reframing them as participants does not, on its own, dismantle deeper systemic forms of oppression:

> Free software, open-source and its equivalents in design—like open design or even earlier versions of critical participatory design—combat userism by empowering users to transcend their roles or avoiding the term "users" altogether, calling them "participants." However, these approaches don't necessarily address other forms of oppression, such as capitalism, racism, sexism, colonialism, or ableism. These systemic issues still contribute to placing people the oppressed in the subordinate position of user even if you treat them as a participant. If you don't address these broader oppressions, you're merely trading one form of oppression for another.

The Role of Technologies in Fostering Engagement

The findings affirm that technology plays a significant and multifaceted role in fostering engagement within PAR (Reason & Bradbury, 2001) and co-design (Sanders & Stappers, 2008). In PAR contexts, the use of technologies, such as apps, online platforms, and devices, was shown to enhance dialogue, mutual understanding, and constructive action among participants, particularly when cultural diversity was acknowledged and addressed. A range of digital tools, including collaborative platforms and multimedia elements (e.g., images and videos), were employed to increase engagement and promote deeper understanding. Technology-aided engagement was also observed in practices such as reflexive use of digital tools and providing feedback on prototypes.

Gibbs et al. (2020) found that digital platforms support flexible data collection and information sharing, offering both flexibility and the potential to enrich data through creative methods. This is consistent with the present findings, which show that various digital tools were used throughout the research process for data collection, sustaining engagement during and between sessions, and facilitating participation. Online platforms also broadened access for those directly affected by the research topic; for example, video conferencing enabled older adults to actively contribute to studies on gerontechnology.

Cole et al. (2022) suggest that higher levels of participation, facilitated by digital engagement, can enhance users' knowledge of technology, support learning, and foster participant empowerment. Similar patterns emerged in the present study, with participants reporting increased technological knowledge, a sense of empowerment, and enjoyment in their participation.

As noted by Tverskoi et al. (2022), social context and attitudes toward technology significantly influence the extent to which digital tools foster engagement in participatory research settings. The findings of the present study further underscore that technology acceptance increases when the target group is actively involved in the development process and meaningfully interacts with the technology. Such involvement ensures that outcomes align with users' real needs and are practical for everyday use.

Limitations

This qualitative study carries limitations typical of interpretive research, where differing conclusions may arise across researchers. To mitigate bias, researcher triangulation was applied (see Appendix 1). The study is also time-bound, and results may have differed if conducted in another period. In terms of generalizability, this study aimed for a comprehensive understanding of the phenomena. To avoid anecdotal results, 32 studies conducted in different countries were analyzed with approximately 5659 participants.

Practical Guidance on Selecting and Using Digital Tools in Research

Building on insights from the studies included, the following guidance summarizes how digital tools can be applied effectively and in which contexts they provide the greatest value for conducting research.

Data collection, workshops, and collaborative analysis. Using videoconferencing platforms like Zoom for real-time interaction, complemented by Mentimeter or similar survey tools to capture both verbal and written input, enhances participation options. Digital whiteboards such as Miro and Ideaflip enable visual organization of ideas, thematic grouping, and live iteration on shared boards, which is particularly valuable for co-design and inclusive engagement. Using digital whiteboards to categorize and organize information is especially effective for structured analysis and decision-making, and they can also support prioritization (e.g., dot-voting) and the co-creation of prototypes.

Maximizing inclusivity through multimodal participation. Platforms including Zoom and Miro facilitate interactive collaboration for groups working across different locations. Providing multiple input channels, such as spoken contributions, written chat, and virtual whiteboards, allows participants to contribute in their preferred format. Combining synchronous methods such as video calls and asynchronous methods such as email or community forums accommodates varying schedules and access needs.

Recommendations for Conducting PAR and Co-design Studies

Drawing on the challenges identified through the thematic analysis of the studies included, we present our recommendations for future PAR and co-design studies.

Strengthening collaborative foundations. The analysis revealed challenges in collaboration and participant engagement. We recommend using technologies such as videoconferencing and online platforms to support relationship building, sustained and active involvement, and broader inclusion. Enhanced accessibility and inclusion through these tools can support meaningful participation and trust-building, ultimately contributing to richer data.

Aligning expectations. The findings revealed instances of unrealistic or mismatched participant expectations. Clear communication of goals, roles, and outcomes is vital for shared understanding. Regularly explaining what is being done, how, and why helps align perspectives, while remote access to meetings supports inclusion and minimizes miscommunication and participant frustration.

Planning for time, funding, and fair compensation. PAR and co-design are often time- and resource-intensive, with limited funding and challenges in fairly compensating participants. The scope and resource demands of PAR and co-design projects can be difficult to anticipate during the grant application stage, as it is challenging to predict participant feedback and the resulting adjustments to plans, timelines, and budgets. Time and funding needs often exceed initial expectations. Ensuring appropriate compensation can further strengthen participant involvement and engagement.

Fostering equitable power-sharing. It is crucial to clearly define decision-making roles and ensure that influence is distributed meaningfully rather than symbolically. Researchers should engage co-researchers as equal partners by balancing structure with openness, letting go of hierarchical norms, questioning assumptions, and recognizing them as valued contributors. While researchers may contribute theoretical expertise, it may not always align with the lived perspectives of co-researchers. We wish to finish with this citation from Devan et al. (2021, p. 1):

> Successful co-design requires a shared commitment and responsibility as a coalition to meet the aspirations of end-users, within the boundaries of time and budget.

Conclusions

PAR served as a guiding framework for co-design, implemented through iterative cycles of planning, action, and review. Both individual and collective reflection were central methodological features, supporting deeper understanding of the issues under investigation. The process also emphasized the development of practical, needs-based solutions.

A range of digital tools, including collaborative platforms and multimedia elements, were employed to enhance participant engagement and foster deeper understanding. Technology-aided engagement was evident in practices such as reflexive use of digital tools and user feedback on design prototypes. These findings reinforce that technology acceptance increases when users are actively involved in development, ensuring outcomes that are relevant, practical, and applicable to everyday contexts.

The appendix to this chapter can be accessed via the designated Google Drive folder. https://drive.google.com/drive/folders/1txlN7pAg41V8Q70aXCqPM5FziwYx7t4n?usp=sharing

References

Armstrong, E., Gapany, D., Maypilama, Ḻ., Bukulatjpi, Y., Fasoli, L., Ireland, S., & Lowell, A. (2022). Räl-manapanmirr ga dhä-manapanmirr – Collaborating and connecting: Creating an educational process and multimedia resources to facilitate intercultural communication. *International Journal of Speech-Language Pathology*, *24*(5), 533–546. https://doi.org/10.1080/17549507.2022.2070670

Benz, C., Scott-Jeffs, W., McKercher, K., Welsh, M., Norman, R., Hendrie, D., Locantro, M., & Robinson, S. (2024). Community-based participatory-research through co-design: Supporting collaboration from all sides of disability. *Research Involvement and Engagement, 10*(1), Article 47. https://doi.org/10.1186/s40900-024-00573-3

Bourgeois, R., Karuri-Sebina, G., & Feukeu, K. E. (2024). The future as a public good: Decolonising the future through anticipatory participatory action research. *Foresight*, *26*(4), 533–549. https://doi.org/10.1108/FS-11-2021-0225

Braun, V., & Clarke, V. (2006). Using thematic analysis in psychology. *Qualitative Research in Psychology*, *3*(2), 77–101. http://eprints.uwe.ac.uk/11735

Bray, E. A., Everett, B., Ajesh, G., Salamanson, Y., & Ramjan, L. M. (2022). Developing a health care transition intervention with young people with spinal cord injuries: Co-design approach. *JMIR Formative Research*, *6*(7), e38616. https://doi.org/10.2196/38616

Bytautas, J., Grigorovich, A., Carson, J., Fowler, J., Goldman, I., Harris, B., Kerr, A., Marcotte, A.-A., O'Doherty, K., Jenkins, A., Kirkland, S., & Kontos, P. (2024). Conversation for change: Engaging older adults as partners in research on gerotechnology. *Research Involvement and Engagement*, *10*(1), Article 27. https://doi.org/10.1186/s40900-024-00557-3

Carter, N., Bryant-Lukosius, D., DiCenso, A., Blythe, J., & Neville, A. J. (2014). The use of triangulation in qualitative research. *Oncology Nursing Forum*, *41*(5), 545–547.

Cech, F., & Wagner, M. (2019). eRollin' on green: A case study on eco-feedback tools for eMobility. In *Proceedings of the 9th International Conference on Communities & Technologies – Transforming Communities*. (pp. 121–125). New York: ACM. https://doi.org/10.1145/3328320.3328402

Chandler, D., & Torbert, B. (2003). Transforming inquiry and action interweaving 27 flavors of action research. *Action Research*, *1*, 133–152.

Chapman, K., Allen, C., & Kendall, E. (2025). Methods for co-designing health communication initiatives with people with disability: A scoping review. *Journal of Health Communication*, *30*(1–3), 120–132. https://doi.org/10.1080/10810730.2025.2462679

Chevalier, J. M., & Buckles, D. J. (2013). *Participatory Action Research: Theory and Methods for Engaged Inquiry (1st ed.)*. London: Routledge. https://doi.org/10.4324/9780203107386

Clark, A. T., Ahmed, I., Metzger, S., Walker, E., & Wylie, R. (2022). Moving from co-design to co-research: Engaging youth participation in guided qualitative inquiry. *International Journal of Qualitative Methods*, *21*, 1–14. https://doi.org/10.1177/16094069221084793

Cockerham, D. (2024). Participatory action research: Building understanding, dialogue, and positive actions in a changing digital environment. *Educational Technology Research and Development*, *72*, 2763–2791. https://doi.org/10.1007/s11423-023-10294-1

Cole, A. C., Adapa, K., Khasawneh, A., Richardson, D. R., & Mazur, L. (2022). Co-design approaches involving older adults in the development of electronic healthcare tools: A systematic review. *BMJ Open*, *12*(7), e058390. https://doi.org/10.1136/bmjopen-2021-058390

Cormick, A., Graham, A., Stevenson, T., Owen, K., O'Donnell, K., & Kelly, J. (2024). Co-designing a health journey mapping resource for culturally safe health care with and for First Nations people. *Australian Journal of Primary Health*, *30*(3), Article PY23172. https://doi.org/10.1071/PY23172

Cortesi, A., Berionni, C., Veeckman, C., Leonardi, C., Schiavo, G., Zancanaro, M., Cescon, M., Sangiuliano, M., Tampakis, D., & Falelakis, M. (2022). Families share: Digital and social innovation for work–life balance. *Digital Policy, Regulation and Governance*, *24*(2), 162–178. https://doi.org/10.1108/DPRG-02-2021-0028

Devan, H., Perry, M. A., Yaghoubi, M., & Hale, L. (2021). 'A coalition of the willing': Experiences of co-designing an online pain management programme (iSelf-help) for people with persistent pain. *Research Involvement and Engagement*, *7*(1), Article 28. https://doi.org/10.1186/s40900-021-00275-0

Fals-Borda, O. (2001). Participatory action research in social theory: Origins and challenges. In P. Reason & H. Bradbury (Eds.). *Handbook of action research: Participatory inquiry and practice* (pp. 27–37). Thousand Oaks, CA: SAGE Publications.

Flicker, S., Maley, O., Ridgley, A., Biscope, S., Lombardo, C., & Skinner, H. A. (2008). e-PAR: Using technology and participatory action research to engage youth in health promotion. *Action Research*, *6*(3), 285–303. https://doi.org/10.1177/1476750307083711

Galabo, R., & Cruickshank, L. (2022). Making it better together: A framework for improving creative engagement tools. *CoDesign*, *18*(4), 503–525. https://doi.org/10.1080/15710882.2021.1912777

Gibbs, L., Kornbluh, M., Marinkovic, K., Bell, S., & Ozer, E. J. (2020). Using technology to scale up youth-led participatory action research: A systematic review. *Journal of Adolescent Health*, *67*(2), S14–S23. https://doi.org/10.1016/j.jadohealth.2019.10.019

Gingell, T., Murray, K., Correa-Velez, I., & Gallegos, D. (2024). A co-design exemplar: How to align with community goals when developing data collection methods with communities from refugee backgrounds. *Research for All*, *8*(1), Article 5. https://doi.org/10.14324/RFA.08.1.05

Hashim, J., Luna Puerta, L., Foong, P. S., Tai, E. S., Yi, H., & Smith, H. E. (2024). Codesigning a digital type 2 diabetes risk communication tool in Singapore: Qualitative participatory action research approach. *JMIR Formative Research, 8*, e50456. https://doi.org/10.2196/50456

Hayes, N., Introna, L. D., & Cass, N. (2021). Participatory design as the temporal flow of coalescing participatory lines. *Computer Supported Cooperative Work (CSCW)*, *30*(4), 507–538. https://doi.org/10.1007/s10606-021-09405-4

Kelly, P. J. (2005). Practical suggestions for community interventions using participatory action research. *Public Health Nursing*, *22*(1), 65–73.

Long, A. F., & Godfrey, M. (2004). An evaluation tool to assess the quality of qualitative research studies. *International Journal of Social Research Methodology*, *7*(2), 181–196.

Macdonald, C. (2012). Understanding participatory action research: A qualitative research methodology option. *Canadian Journal of Action Research, 13*, 34–50. https://doi.org/10.33524/cjar.v13i2.37

McIntyre, A. (2008). *Participatory Action Research*. Thousand Oaks, CA: SAGE Publications. https://doi.org/10.4135/9781483385679

Moghimi, E., Belfry, K., Farr, S., Stafford, S., Bogdan, A., Brush, M., Canning, C., & Kim, S. (2025). Using a co-design approach to develop a preventative online mental health program for youth (POMHPY): A quality improvement project. *BMC Health Services Research, 25*(1), Article 219. https://doi.org/10.1186/s12913-024-12101-w

Molapo, M., Densmore, M., & Morie, L. (2016). Designing with community health workers: Enabling productive participation through exploration. In *Proceedings of the First African Conference on Human Computer Interaction, AfriCHI 2016: Building Bridges, Breaking Barriers, 21-25-,* 58–68. https://doi.org/10.1145/2998581.2998589

Munn, Z., Peters, M. D. J., Stern, C., Tufanaru, C., McArthur, A., & Aromataris, E. (2018). Systematic review or scoping review? Guidance for authors when choosing between a systematic or scoping review approach. *BMC Medical Research Methodology, 18*, 143. https://doi.org/10.1186/s12874-018-0611-x

Onwezen, M. C., Bouwman, E. P., & van Trijp, H. C. M. (2021). Participatory methods in food behaviour research: A framework showing advantages and disadvantages of various methods. *Foods*, *10*(2), 1–19. https://doi.org/10.3390/foods10020470

Paluch, R., Cerna, K., Kirschsieper, D., & Müller, C. (2023). Practices of care in participatory design with older adults during the COVID-19 pandemic: Digitally mediated study. *Journal of Medical Internet Research, 25*, e45750. https://doi.org/10.2196/45750

Perry, M. A., Devan, H., Davies, C., Hempel, D., Ingham, T., Jones, B., Reid, S., Saipe, B., Patient Advisory Group Members, Technology Design Team, & Hale, L. (2022). iSelf-help: A co-designed, culturally appropriate, online pain management programme in Aotearoa. *Research Involvement and Engagement*, *8*(1), Article 6. https://doi.org/10.1186/s40900-022-00339-9

Reason, P., & Bradbury, H. (Eds.). (2001). *Handbook of Action Research: Participative Inquiry and Practice*. London: SAGE Publications.

Redublo, T., Paul, S., Joshi, A., Arbour, S., Murray, R., & Chiu, M. (2024). We-care-well: Exploring the personal recovery of mental health caregivers through participatory action research. *Frontiers in Public Health, 12*, Article 1366144. https://doi.org/10.3389/fpubh.2024.1366144

Saldaña, J. (2013). *The Coding Manual for Qualitative Researchers*. New York: SAGE Publications.

Sanders, E. B.-N., & Stappers, P. J. (2008). Co-creation and the new landscapes of design. *CoDesign*, *4*(1), 5–18. https://doi.org/10.1080/15710880701875068

Steen, M. (2013). Co-design as a process of joint inquiry and imagination. *Design Issues*, *29*(2), 16–28. https://doi.org/10.1162/DESI_a_00207

Stringer, E. (1996). *Action Research: A Handbook for Practitioners*. Thousand Oaks, CA: SAGE Publications.

Taba, M., Ayre, J., McCaffery, K., Vassilenko, D., Ma, I. C. K., Haynes, T., Leask, J., Wilson, A., & Bonner, C. (2025). Co-designing a framework for social media health communication to young people: A participatory research study. *Health Expectations*, *28*(2), Article e70203. https://doi.org/10.1111/hex.70203

Tariq, S., Grewal, E. K., Booth, R., Nat, B., Ka-Caleni, T., Larsen, M., Lawson, J., Whaley, A., Walsh, C. A., & Campbell, D. J. T. (2023). Lessons learned from a virtual community-based participatory research project: Prioritizing needs of people who have diabetes and experiences of homelessness to co-design a participatory action project. *Research Involvement and Engagement*, *9*(1), 46. https://doi.org/10.1186/s40900-023-00456-z

Treasure-Jones, T., & Joynes, V. (2017). Co-design of technology-enhanced learning resources. *The Clinical Teacher*, *15*(4), 281–286. https://doi.org/10.1111/tct.12733

Trischler, J., Dietrich, T., & Rundle-Thiele, S. (2019). Co-design: From expert- to user-driven ideas in public service design. *Public Management Review*, *21*(11), 1595–1619. https://doi.org/10.1080/14719037.2019.1619810

Tverskoi, D., Babu, S., & Gavrilets, S. (2022). The spread of technological innovations: Effects of psychology, culture and policy interventions. *Royal Society Open Science*, *9*(6), Article 211833. https://doi.org/10.1098/rsos.211833

van Amstel, F., Serpa, B., & Secomandi, F. (2025). Systemic oppression in service design. In M. Suoheimo, P. Jones, S.-H. Lee, & B. Sevaldson (Eds.). *Systemic Service Design: Design for Social Responsibility* (pp. 106–124). Abingdon: Routledge. https://doi.org/10.4324/9781003501039-7

van Amstel, F. M. C. (2024, November 25). *Systemic userism in service design* [Lecture]. TU Delft, Delft, The Netherlands. https://fredvanamstel.com/talks/systemic-userism-in-service-design

van Leersum, C. M., Jaschinski, C., Bults, M., & van der Zwart, J. (2024). Citizen involvement in research on technological innovations for health, care or well-being: A scoping review. *Health Research Policy and Systems*, *22*(1), Article 119. https://doi.org/10.1186/s12961-024-01152-4

Vargas, C., Whelan, J., Brimblecombe, J., & Allender, S. (2022). Co-creation, co-design, co-production for public health: A perspective on definition and distinctions. *Public Health Research & Practice*, *32*(2), e3222211. https://doi.org/10.17061/phrp3222211

Wallis, P. J., Bosomworth, K., Harwood, A., & Leith, P. (2017). Charting the emergence of a 'knowing system' for climate change adaptation in Australian regional natural resource management. *Geoforum*, *84*, 42–50. https://doi.org/10.1016/j.geoforum.2017.06.002

Webster, J. & Watson, R. T. (2002). Analyzing the past to prepare for the future: Writing a literature review. *MIS Quarterly*, *26*, 13–23.

Zamenopoulos, T., & Alexiou, K. (2018). *Co-Design as Collaborative Research*. University of Bristol/AHRC Connected Communities Programme.

Zhu, R., Hardy, D., & Myers, T. (2019). Co-designing with adolescents with autism spectrum disorder: From ideation to implementation. In *Proceedings of the 31st Australian Conference on Human-Computer Interaction* (pp. 106–116). New York: ACM. https://doi.org/10.1145/3369457.3370914

3 Living Law Labs as Public Sites of Participatory AI

Experimental Infrastructures for Ethical AI Futures

Jules Rochielle Sievert, Yue Ang, and Miso Kim

Introduction

This chapter takes an exploratory and argumentative approach to critically examine the conceptual and infrastructural possibilities of Living Law Labs. The distribution of power across legal and technological systems is not neutral; it has historically reinforced systemic inequities by privileging certain groups while marginalizing others. Law and technology do not operate independently. Instead, they function as entangled systems that shape how human, algorithmic, and environmental agents interact. These architectures shape who can participate in civic life, access essential resources, and benefit from technological and ecological advancements, while simultaneously determining who is excluded.

Recognizing the entanglement of law and technology requires transdisciplinary inquiry into how power, agency, and access are co-constructed through socio-technical systems. While design offers critical tools for surfacing these dynamics, institutional and epistemic silos often obscure how harms such as algorithmic bias and digital exclusion are materially produced through the overlapping infrastructures of law, technology, and society. To address this critical gap in scholarship and practice, we introduce the Living Law Lab: a participatory, practice-based framework for prototyping ethical and community-rooted approaches to AI governance. As legal designers, educators, and practitioners, we argue that law and design are deeply interconnected, with design serving as the interface through which law becomes legible, accessible, and actionable in society.

By addressing the need to distinguish between law and justice, we propose a new model of community-led AI governance rooted in transdisciplinarity and informed by the lived experiences of marginalized communities. In the face of increasingly complex socio-technical systems and persistent structural injustice, siloed approaches to law, design, and technology are inadequate. A shift is needed from top-down, expert-driven models toward relational, integrated, and community-embedded practices. While legal design has improved usability and access, it often fails to confront the deeper power imbalances that legal systems continue to reproduce.

While law schools and related institutions continue to be shaped by professional norms, accreditation mandates, and short-term service delivery models, experimental shifts are beginning to emerge. These shifts signal a growing dissatisfaction with transactional approaches to legal aid and a desire for more profound, more systemic change. Living Law Labs respond to this moment as a radical reorientation toward long-term, community-led co-governance.

The Living Law Lab emerges in response to these limitations of siloed thinking and transactional acts, offering a more expansive and integrated approach. It builds on these educational models but reconfigures them within a transdisciplinary pedagogical framework, bringing together law, design, technology, and critical social theory. This approach is not only about

DOI: 10.4324/9781003667872-4

understanding the complexity of socio-technical systems but also about cultivating the ability to intervene in them meaningfully and ethically.

Unlike traditional legal education settings, the Living Law Lab is not merely a site of instruction or service provision. It is an applied infrastructure for ethical investigation and civic co-creation. Through community co-design, participatory research, AI audits, and rights-based public education campaigns, the Lab supports communities in actively reshaping the legal and technological systems that govern their lives. This environment cultivates the capacity of both practitioners and community members to move beyond critique toward collective action and sustained engagement. It supports a deeper understanding of the temporal life of data, exploring how its collection, circulation, and use over time impact the rights of individuals and communities historically marginalized and excluded from participating in data governance.

Through gap-spotting and its deficiencies, the Living Law Lab addresses the need identified by Alvesson and Sandberg (2011) by moving beyond gap-spotting to problematize core assumptions in law and AI governance. Rather than treating public participation as supplemental, it redefines justice as co-created through participatory, transdisciplinary infrastructures. The model challenges the legitimacy of professional gatekeeping and the neutrality of technical systems, offering a practice-based response to Alvesson and Sandberg's call for research that disrupts dominant paradigms and reimagines foundational concepts.

The following section explores what becomes possible when space is created for genuine transdisciplinary collaboration for public engagement. Through the Living Law Lab model, law, design, technology, and critical social theory converge to cultivate new civic capacities, knowledge practices, and modes of governance. Positioned as experimental, community-led infrastructures, Living Law Labs demonstrate how transdisciplinarity can enable more just and responsive systems by confronting structural injustice across time, space, and lived experience. This approach moves beyond disciplinary reform toward the co-creation of shared futures shaped by those most affected.

Legal Clinics to Civic Co-creation: Reimagining Experiential Legal Education for Participatory Futures

Legal education over the years has integrated the practical elements of legal practice into the legal curriculum (Giles & Ang, 2025). Law schools now offer law clinics, legal advice centers, *pro bono* opportunities, and law placements to enrich well-rounded legal practitioners. Law clinics and legal advice centers are part of a similar branch of services that law schools provide. Under the supervision of practicing solicitors, students gain experience through client interviewing, legal drafting, and giving client advice.

Pro bono (Pro Bono Week, 2021) and law placements are in a similar branch of legal experience where students go to legal service centers and work with the firm. They will gain similar legal experience with their legal advice center and law clinic counterparts, with more demands on professionalism. These firms are often located off campus and operate in a corporate setting. This constitutes '*hands-on legal work*' that holds significant experiential value for students seeking to become practice-ready. These opportunities are optional, offering students a chance to build practical skills outside the required law curriculum. As such, they tend to attract those who are self-motivated to pursue experiential learning alongside their legal studies.

In the increasingly difficult climate where the cost of living is diverting students' attention to focus on studies and away from extracurricular activities (Simkovic & McIntyre, 2014), some law schools have increasingly incorporated these extracurricular activities into the law curriculum (Giles & Ang, 2025). For example, law programs with a named pathway, such as undergraduate law with placement or postgraduate law with the solicitor qualifying exam (SQE). The prescribed inclusion of the legal practice exposure into the law curriculum appeals to those who would like to have a head start in the legal profession. Law schools that center engagement with the local community offer street law and public legal education.

Law schools that offer street law, programmatic, and coursework in legal design or public legal education serve as a bridge that is meant for outreach, public engagement, and educating citizens about rights and responsibilities. Street law (Blackburn, 2023) and public legal education (Drumbl, 2023) are within a similar branch, with an element of public engagement and an element of educating the selected public about their rights. These activities can either run alongside the legal curriculum or be incorporated into the legal curriculum as an option. Students who are involved in these activities are encouraged to learn about legal rights that affect a selection of the public, with which they will engage, and design a short public engagement event that educates them. We take the inspirations of street law and public legal education and advance these concepts into a Living Law Lab.

The Living Law Lab model reimagines both the application of law and the purpose of legal education. Instead of training students solely to become practice-ready professionals who operate within existing systems, the model cultivates civic-ready collaborators—individuals equipped to work across disciplines and alongside communities to challenge, redesign, and transform those systems.

Rooted in a transdisciplinary, community-centered framework, the Lab situates legal education not as a pipeline to professionalization but places a special emphasis on public engagement as a site of democratic participation, ethical inquiry, and structural intervention. It invites law students, designers, technologists, and community members into shared projects that blur the boundaries between legal knowledge, civic agency, and public innovation.

An example of a public engagement activity is law students educating school children and college students about youth violence and knife crime. London South Bank University offers this activity in two variations: street law and public legal education. The approach of street law is a collaboration between university law students and college and school students in understanding everyday legal matters. Street law is an interaction where participants educate school and college students through a mock trial. Such a performance enriches the school and college students' understanding of the criminal justice system and the perpetrators and victims of knife crime. The skills developed through these types of experiential engagements help broaden the worldview and foster a deeper understanding of advocacy and social justice issues.

Public legal education is more about academic interaction with school and college students. This option is embedded in the law curriculum. Law students will have mastered the legal concepts of youth violence and knife crime as an issue within the locality of the selected schools and colleges. They design an engagement event in which they have complete choice of its delivery to perform in the schools and colleges in the locality and educate the audiences about citizens' rights and their role in knife crime. Such an interaction communicates awareness of the severity of knife crimes and the role each student can play as a key agent in tackling these issues. In both instances, the collaboration and creation of the law students do not need to be supervised by a legal practitioner. These activities closely align with the understanding of law labs as opposed to law clinics or legal advice centers. These activities are more in the scope of legal education than legal practice.

As law schools increasingly emphasize the development of practice-ready skills alongside legal knowledge and specialization, expanding opportunities for meaningful interaction between students and the public has become essential. Moreover, law firms are expecting law graduates to demonstrate AI literacy. There is a shift in the expectation of becoming practice-ready in an age where lawyers are experimenting with AI-driven technologies for service delivery. Initiatives such as law school clinics, public legal education, and street law programs have laid vital groundwork for experiential learning by connecting students with real-world legal contexts.

However, these models also reveal a deeper problem within legal training as addressed in the form of AI literacy. They are often shaped by disciplinary silos, institutional norms, and service-based frameworks that often prioritize casework over systems thinking. They frequently position the public as recipients of legal knowledge rather than as co-creators of legal meaning or designers of governance. Moreover, they rarely prepare students to engage with the broader socio-technical systems.

Living Lab as a Design Approach

Living Labs are participatory environments where communities help shape the systems that govern their lives through iterative, relational engagement. Emerging from earlier innovation movements, Schuurman et al. (2025) trace their origins to three foundational predecessors: the Scandinavian Cooperative and Participatory Design Movement of the 1960s and 1970s, which emphasized collaborative design in workplace settings; European Social Experiments with IT in the 1980s, which focused on leveraging technology for community benefit; and the Digital City Projects of the 1990s, which brought together public, private, and civic actors to explore the role of digital platforms in urban development (pp. 24–25).

MIT's Professor William Mitchell first coined "*Living Lab*" to describe experimental home-like research settings. For example, Kidd *et al* in the late 1990s conducted a living home laboratory for computing research. The living home lab consisted of two apartments, one for control and the other with technology for human-centered technology assistance in everyday living (Kidd et al., 1999). Living Labs have since become a global movement (Leminen & Westerlund, 2019). However, the European approach diverged by embedding innovation directly in users' real-life environments, leading to the formation of the European Network of Living Labs (ENoLL) in 2006 during Finland's EU Presidency (Schuurman et al., 2025, pp. 26–27). Living Labs are defined as "user-centered", open innovation ecosystems based on a systematic user co-creation approach, integrating research and innovation processes in real-life communities and settings (Schuurman et al., 2025, p. 18). They operate within the "*Quadruple Helix*" model, which incorporates public authorities, private enterprises, academia, and civil society into innovation processes (Schuurman et al., 2025, pp. 18–19). Methodologically, Living Labs uses a multi-method, problem-driven approach tailored to each challenge. This ensures methodological fit-for-purpose across exploratory and confirmatory processes (p. 20).

This model helps to build in a range of diverse perspectives that can contribute to shaping the process so that solutions and interactions can be more holistically designed and implemented. This model represents a significant evolution in innovation ecosystems by integrating citizens, civic organizations, and NGOs as central actors in the co-creation process. Extending the traditional triple helix framework, which focuses on collaboration between academia, industry, and government, the quadruple helix emphasizes inclusive, participatory approaches that reflect diverse societal perspectives (Harbers & Overdiek, 2022, p. 6). By involving citizens and communities directly in the ideation and development of new technologies, Living Labs grounded in the quadruple helix allow for horizontal co-creation, where stakeholders are not merely consulted but are empowered to help define problems and co-design solutions (Harbers & Overdiek, 2022, p. 7).

By integrating co-creation, real-life testing, stakeholder diversity, and methodological adaptability, the Living Lab model provides a framework for socially grounded and impactful innovation. This approach aligns with Taylor et al.'s (2025) call for justice-centered, community-driven infrastructures capable of illuminating the lifecycle of data and exposing the structural nature of algorithmic harm. They document how data collected without consent is often repurposed in ways that reinforce profiling, exclusion, and automated governance, typically without the knowledge or influence of those most affected (Taylor et al., 2025, pp. 6–7). These harms, they argue, are cumulative and resistant to redress under traditional legal or regulatory mechanisms.

The Living Law Lab addresses this challenge by cultivating community capacity to trace, interrogate, and reimagine these data lifecycles. In doing so, it embodies the kind of infrastructural accountability Taylor et al. (2025) envision: one that enables affected communities to build counter-narratives, demand transparency, and shape redress mechanisms rooted in justice (p. 11). As Schuurman et al. (2025) demonstrate, Living Labs represent a shift toward more democratic and context-sensitive innovation systems that align closely with societal needs. The authors describe how policy-focused Living Labs, particularly Urban Living Labs and Regulatory Sandboxes, provide platforms for public authorities and stakeholders to collaboratively explore and evaluate legal frameworks and policy innovations tailored to local contexts (p. 42).

Additionally, Living Labs are increasingly engaging with the ethical and legal implications of digital transformation. ENoLL has formed thematic working groups, such as "Social Innovation and Digital Rights" and "Gender in Innovation," to promote equity, justice, and inclusion in developing emerging technologies. These efforts underscore the role of Living Labs in advancing responsible innovation aligned with societal values (Schuurman et al., 2025, pp. 44–45).

Building on this foundation, the Living Law Lab offers a practice-based model for critical inquiry and design at the intersection of law, technology, and community power. Situated within the overlapping fields of legal design, public interest technology, and community organizing, the Lab operates as a dynamic ecosystem of practices, relationships, and shared inquiry. It advances a transdisciplinary pedagogy that reimagines law, design, and AI governance as interconnected and participatory processes, grounded in collective sense-making, critical resistance, and the cultivation of civic literacy attuned to structural injustice.

Pushing the Boundaries Within the Lines

Translating the Living Law Lab from concept to practice requires careful attention to legal and ethical boundaries, particularly the regulatory constraints surrounding the unauthorized practice of law (UPL). In the United States, these constraints pose significant challenges for interdisciplinary, community-based legal innovation. While the Lab encourages collaboration across legal, design, and civic domains, it must do so without compromising professional standards or violating jurisdictional rules governing legal practice.

Chien and Kim (2025), in *Generative AI and Legal Aid: Results from a Field Study and 100 Use Cases to Bridge the Access to Justice Gap*, frame UPL as a double-edged constraint intended to protect consumers from unqualified legal assistance but often acting as a structural barrier to innovation. They argue that fears of UPL violations can inhibit partnerships, suppress investment, and stall the development of scalable, AI-enabled legal aid tools, particularly in resource-constrained public interest settings (pp. 918–920). The authors also critique the enforcement of UPL statutes as, at times, serving more as professional protectionism than as genuine public safeguarding.

Daryl Lim (2025) situates this tension within broader debates about how to govern emerging technologies responsibly. Comparing global models, Lim notes that the European Union takes a precautionary, rights-based approach to AI governance. At the same time, the United States adopts a sectoral, principles-based regime that delegates authority to agencies such as the Federal Trade Commission (p. 186). Each model presents trade-offs: overregulation may stifle innovation, especially for small and mid-sized developers, whereas underregulation risks entrenching algorithmic bias and systemic harm. Lim ultimately advocates for "regulatory pluralism"—a layered model that combines hard law with soft law tools such as voluntary standards and ethical codes to guide responsible innovation (p. 187).

Crucially, Lim argues that regulatory design alone is insufficient. Legitimacy in AI governance hinges on building public trust through inclusive, participatory processes. "*Trust*," he writes, "*is not merely a corollary of compliance but a key determinant of social license and legitimacy*" (p. 194). Effective governance must include historically marginalized communities not just as stakeholders to be consulted, but as co-designers with shared decision-making power. This requires sustained investment in digital and civic literacy, as well as in institutional mechanisms that resist regulatory capture by elite or corporate interests (Lim, 2025, p. 195).

The Living Law Lab model emerges in response to these intersecting tensions. Rather than treating UPL as a fixed barrier, the Lab approaches it as a productive design constraint, one that shapes how participation, accountability, and expertise are structured. As a participatory platform, the Living Law Lab offers a generative civic infrastructure for critical engagement. It creates space for communities to examine the policies, systems, and algorithmic processes that shape their lives. Through collective inquiry, participants uncover how AI systems operate, surface exclusions embedded in legal and technical infrastructures, and co-develop community-driven interventions in law and policy.

By focusing on public legal education, participatory policy development, and systemic design rather than individualized legal representation, the Living Law Lab operates largely outside the boundaries of UPL regulation. In accordance with the American Bar Association (ABA, 2023) Model Rule 5.5, the Lab avoids unauthorized practice by ensuring that any legal tasks are carried out under the supervision of licensed attorneys or legal aid organizations and that its core activities fall within permitted domains of public education and advocacy. The ABA permits nonlawyers to engage in legal information sharing and public-facing legal education, provided they do not offer specific advice or services to individual clients (American Bar Association, 2023).

The model serves as both a site of legal empowerment and a safeguard against regulatory overreach, aligning with calls for participatory, ethically supervised AI governance (Chien & Kim, 2025; Lim, 2025). But its transformative potential lies beyond compliance—in designing for public trust by reimagining AI as a participatory medium for co-creating transparency, accountability, and justice. This shift marks a move from regulatory accommodation toward a design justice paradigm.

Public Trust and Design

Trust becomes a central axis in this reorientation, serving both as a precondition for participation and as an outcome of participatory design. As AI systems increasingly shape daily life, the structural parallels between legal and technological infrastructures become evident. Both can reproduce systemic oppression or to be reimagined as tools for collective liberation. Karpen and Senova (2021), in *Designing for Trust: Role and Benefits of Human-Centered Design in the Legal System*, explore how human-centered design (HCD) can help build trust in legal institutions, particularly courts. They argue that trust is foundational to the effectiveness of legal systems, especially for lay people who often face emotional, psychological, or cultural barriers when engaging with legal institutions. Legal environments that fail to account for these human experiences risk alienating court users, diminishing their willingness to engage or accept legal outcomes (Karpen & Senova, 2021, Introduction).

The authors define trust as a willingness to accept vulnerability based on positive expectations of others (Karpen & Senova, 2021, Introduction). In the legal context, this manifests when users feel safe, informed, and supported throughout their interaction with the court. HCD offers a systemic approach to crafting such experiences. It prioritizes human needs and emphasizes empathy, participation, iterative learning, visualization, and systems thinking (Karpen & Senova, 2021, Section 2).

They introduce five levers—place, process, people, purpose, and planet—as design focal points and identify two levels of design: interactional (e.g., signage, waiting rooms, digital platforms) and institutional (e.g., culture, norms, unspoken rules) (Karpen & Senova, 2021, Sections 4.1–4.2). These elements collectively form a framework courts can use to support trust-building experiences throughout the user journey, not just in the courtroom. By engaging court users and stakeholders in participatory design processes, courts can become more just, accessible, and humane (Karpen & Senova, 2021, Section 4.3).

However, as AI technologies become increasingly embedded in the fabric of daily life, public trust has emerged as a central concern, raising urgent questions about who these systems are built for, who is excluded, and how power dynamics shape their development. The Artificial Intelligence Index Report (2024), produced by Stanford's Human-Centered AI Institute, highlights this issue through a comprehensive examination of trust in AI systems. The report underscores that responsible, transparent communication is essential for meaningful public engagement and societal readiness. As AI continues to influence social, political, and economic domains, public trust depends not only on regulatory safeguards but also on how accessible, explainable, and accountable both the technologies and their developers are (Artificial Intelligence Index Report, 2024, p. 163). While dominant discourses on AI trust often emphasize transparency and explainability, these technical metrics overlook the deeper structural conditions that shape public confidence. Participatory approaches must move beyond communication strategies and consider how design itself can reinforce or resist systemic exclusion.

In User Oppression in Human-Computer Interaction: A Dialectical-Existential Perspective, Tors Pedro Ferreira, José Abdelnour-Nocera, and Hugo G. Nicolau argue that systemic oppression in (HCI) reduces opportunities for users to "become-more" by limiting their agency through design practices that reinforce existing hierarchies, rather than empowering users to actively engage with designing technology, confining them to passive roles and perpetuating a "user-as-object" paradigm. (Gonzatto & van Amstel, 2022, p. 764). The authors critique "userism" as an ideological framework that perpetuates inequalities by positioning users as passive recipients of technological progress rather than active participants in its development. This framework reinforces a hierarchical structure where designers maintain control, limiting the *"end-users"* ability to influence or shape systemic design practices. This dynamic dehumanizes both designers and users: designers are confined to the role of authority figures, while users remain constrained in their capacity for creative and transformative engagement.

These critiques of human-computer interaction parallel and deepen broader concerns scholars raise about the societal impacts of AI systems more broadly. Like Ferreira, Abdelnour-Nocera, and Nicolau, who critique the "*user-as-object*" paradigm in HCI, other scholars have illuminated how AI technologies are embedded within and shaped by existing power structures. This convergence of scholarship underscores that the problems of limited user agency, hierarchical design practices, and systemic exclusion are not isolated to HCI but are symptomatic of more significant socio-technical dynamics.

Participatory Power Sharing

Building on this, Smith (2025) warns that although AI has the potential to contribute $2.6 trillion to $4.4 trillion to the global economy annually (Smith, 2025). Without intervention, the risk of job displacement may increase due to automation across various sectors (Smith, 2025). In the United States, Black Americans are 10% more likely to be in roles susceptible to AI-driven automation, potentially affecting 4.5 million jobs (Smith, 2025). Beyond economic displacement, AI bias in training data has reinforced discrimination in hiring, lending, housing, and compensation practices (Smith, 2025). Wealthier nations and corporations are best positioned to adopt AI, widening the digital divide and global economic disparities (Smith, 2025). To prevent AI from becoming a tool that entrenches inequality, Smith (2025) argues for a universal commitment to AI accessibility and digital inclusion.

Puzio (2025) argues that technologies do not just represent individuals but also actively shape conceptions of humanity. She contends that definitions of "human" and "non-human" are constructed through both physical and social processes, shaped by political, institutional, and technological influences (p. 2348). The development of technologies involves contributions from designers, developers, regulatory frameworks, and political actors, resulting in specific versions of bodies and identities. This process embeds power relations within technological systems (p. 2348).

Puzio (2025) supports her argument by referencing airport body scanners, which classify transgender individuals as threats when their bodies do not conform to "pre-programmed algorithmically identified male or female gender shapes" (p. 2349). She also observes that facial recognition technologies "fail to recognize black faces" (p. 2349). These examples indicate that technological systems do not objectively detect bodies; instead, they operate based on embedded assumptions regarding gender and race.

According to Puzio, these examples are significant for what they reveal about the production of knowledge. Technologies materialize specific conceptions of the human, while rendering other bodies less intelligible or less worthy of respect. Through these cases, she demonstrates that definitions of the human and non-human are closely linked to power, and that technological design plays a role in determining who is recognized, categorized, and valued (p. 2348).

Like Puzio and Smith, Miriam Tedeschi (2024) explains how smart city technologies generate ethical issues relating to datafication, surveillance, geosurveillance, profiling, social sorting, and anticipatory governance, arguing that these processes have significant consequences that may reproduce and reinforce inequalities that impact how individuals are viewed and treated (Tedeschi 2024, pp. 1–13). Tedeschi situates these developments within the context of neoliberal and cybernetic urbanism, in which automated extraction of personal data becomes constitutive of urban control and security, and in which data functions as capital inside a political–economic regime (pp. 1–13).

Tedeschi conceptualizes datafication as the transformation of social and spatial life into continuously harvested socio-economic and locational data, producing "data-doubles" that circulate within digital infrastructures over which individuals have no meaningful control (pp. 1–13). These infrastructures do not simply overlay the city but increasingly participate in its production, shaping access to space, mobility, inclusion, and everyday urban experience (pp. 1–13).

She further argues that practices such as predictive policing, automated welfare decision-making, algorithmic classification of neighbourhoods, and live monitoring operate through anticipatory governance logics that pre-emptively identify and manage risk (pp. 1–13). While framed as optimising efficiency, safety, and optimisation, these systems operate through anticipatory logics that construct and govern risk, reproducing socio-spatial exclusion, redlining, and discrimination, and thereby generating new forms of urban (in)justice (Tedeschi 2024, pp. 1–13).

Taken together, the work of Puzio, Smith, and Tedeschi exposes a layered and compounding set of exclusions embedded within the design, deployment, and governance of AI and digital infrastructures. These authors reveal how various technologies often reproduce and intensify structural inequalities through processes like datafication, automation, and predictive governance, revealing a paradox faced by those most impacted by these systems.

In *Participatory AI: Forging Shared Frameworks for Action,* Davies, Dencik, and Jones argue that participatory design in artificial intelligence must move beyond superficial engagement and aim to redistribute power within governance systems. They emphasize that authentic participation requires institutional mechanisms that enable communities to shape how AI technologies are developed, used, and regulated. The authors introduce two models to support this shift. The first, called the *AI Issue Association* model, focuses on creating structured, multi-stakeholder groups organized around specific AI-related concerns. These groups are intended to transform public input into actionable decisions by grounding participation in concrete issues. Rather than relying on vague or symbolic engagement practices, this model institutionalizes participation through formal governance channels that enable meaningful public influence over AI outcomes (Davies et al., 2025).

The second model, *Digital Self-Determination*, prioritizes the agency of communities that have been historically excluded from technology governance. Drawing on postcolonial and cultural sovereignty theories, this model calls for the design of digital systems that are directly accountable to those most affected by them. It proposes co-governance arrangements that include shared authority, long-term community involvement, and control over digital infrastructures. Participation, in this view, is not limited to consultation but must lead to shifts in power and ownership (Davies et al., 2025). Together, these two models challenge traditional approaches to AI governance by calling for participatory structures that embed democratic principles into the development and oversight of AI. Davies, Dencik, and Jones contend that participatory AI must be treated as a form of institutional power sharing rather than a procedural or symbolic gesture.

These patterns reflect what Ferreira, Abdelnour-Nocera, and Nicolau describe as the "user-as-object" paradigm, in which people are treated as passive recipients rather than co-designers of technology (Gonzatto & van Amstel, 2022, p. 764). In contrast, participatory justice rejects this dynamic. It reframes design as a practice of sharing power, not merely collecting input. Centering those most affected by socio-technical systems, it calls for structural engagement embedded within institutions. The Living Law Lab builds on this approach by creating community-led, transdisciplinary spaces where law and technology are not only co-designed but reshaped in relation to one another, challenging their traditional boundaries and forging new methods of critical inquiry, knowledge building, resistance, and repair.

From Vision to Practice

So, how do we move from theory to practice? Transformation requires more than critique; it demands sustained action across pedagogy, design, and policy. While diagnosing the harms embedded in legal and technological systems is essential, meaningful change begins when theory is translated into participatory practice: through educational spaces that cultivate civic agency, design processes that redistribute power, and policy dialogues grounded in community experience.

The Living Law Lab model offers a generative terrain for imagining, testing, and co-creating legal and technological futures as contested social, cultural, and political possibilities shaped by values, governance choices, and power relations. In this framing, technology is never neutral: it can reproduce existing inequities or serve as a site of resistance, repair, and transformation. Living Law Labs create the conditions for various communities to critically question, shape, and reimagine these futures so they reflect collective priorities and justice-centered outcomes. As spaces of practice, these labs enable justice-based experimentation by making visible the tensions, frictions, and negotiations inherent in collaborative work. They prioritize relational, cultural, and political infrastructures, trust, accountability, mutual recognition, and civic agency so that governance, design, and participation remain context responsive.

We have been laying the groundwork for the Living Law Lab through the East Boston Spatial Justice Lab by developing community-driven legal design practices, building relationships with local partners, and piloting participatory methods that center lived experience, legal empowerment, and spatial justice. Over the past years, we have come to recognize this work as a transdisciplinary and transformative practice, one that draws from, integrates, and moves beyond disciplinary boundaries to address complex systemic issues. It operates through sustained academic–community alliances, bringing together law schools, design researchers, advocacy groups, and grassroots organizations in shared inquiry and action. This model is inherently scalable, capable of moving across disciplines, geographies, and policy contexts to support more just and community-responsive legal futures.

Building a Living Law Lab begins with the recognition that relationships are infrastructure. Establishing that infrastructure starts by mapping interconnected social ecosystems across legal, cultural, technological, academic, and grassroots spheres to identify potential partners, uncover existing capacities, and understand the networks already shaping the local landscape. This early work is central to developing trust and values alignment, ensuring that future collaboration is grounded in shared priorities and community-based learning. Community members, advocates, researchers, and designers work together from the outset to frame questions and co-create knowledge, linking lived experience to broader goals for systemic change.

The lab grows through a broad network of partnerships that connect local participants with legal clinics, advocacy organizations, technologists, and policy actors. These connections strengthen both capacity and impact, allowing the lab to serve as a hub for local collaboration and a bridge to resources, expertise, and opportunities beyond the immediate geography. Open-source tools, shared learning platforms, and peer-to-peer exchanges keep the lab flexible, adaptive, and scalable while ensuring it remains rooted in local relationships.

Using methods drawn from participatory action research and co-design, the lab hosts activities such as collaborative design sessions, skill-sharing workshops, community dialogues, participatory mapping, and collective learning exchanges. These are structured as iterative cycles of inquiry, creation, and reflection, positioning community members as co-researchers and co-designers. This approach enables participants to define and test interventions, evaluate results, and refine strategies in real time, integrating education, design, and policy development into a collaborative practice that produces outputs both grounded in lived experience and built for lasting impact.

The process begins with a small series of pilot conversations where community members, advocates, academics, and designers surface lived experiences, identify pressing challenges, and shape ideas for joint action. From there, the lab develops targeted participatory projects that can expand over time without losing their grounding in trust, shared values, and community priorities.

Conclusion

Law and AI are parallel architectures of power, distinct in form, yet deeply aligned in how they shape rights, structure society, and determine who is visible, valued, and heard. Though often framed as neutral or inevitable, both are built atop histories of exclusion. They are infrastructures that can either codify inequality or be reimagined as tools for care and accountability. Both are designed and maintained, which means they can also be redesigned and critiqued.

When unexamined, law can legitimize the harms produced by AI. Reimagined, it can become a site of resistance, a means to interrogate algorithmic values, challenge opaque decisions, and advocate for accountability. These infrastructures often operate in mutually reinforcing ways, each shaping and validating the other's logics, structures, and exclusions. Their entanglement demands critical examination and profound transformation, one that goes beyond regulating AI through law to co-creating socio-technical systems rooted in justice, transparency, and community power. Through this shift, we move from inherited harm toward futures shaped by collective imagination and participation.

The Living Law Lab serves as a site and model for deep observation, dialogue, and transformative pedagogy. It offers an approach that can reveal how exclusionary systems take shape over time and how they can be challenged through participatory, community-anchored infrastructures. As an educational space, the Lab engages with the temporal life of data, tracing how historical injustices are encoded into present systems and projected into future constraints. Within this framework, justice is not a fixed outcome but an evolving, intergenerational practice grounded in care, critical inquiry, and mutual accountability.

This moment requires deep reflection regarding how we govern, educate, and design. New pedagogies are needed, ones that dismantle disciplinary silos and prepare students to interrogate power in law, technology, and design. Legal training alone cannot anticipate the normative stakes of AI; design education alone cannot repair systemic harm. But together, they can cultivate the critical capacities and explore alternatives grounded in equity and civic imagination, offering both a framework and a practice, an open invitation to co-create systems that do not merely reflect the world as it is but help shape it into what it must become.

The Living Law Lab offers more than a theoretical intervention; it is a replicable, adaptable framework for building participatory, justice-oriented infrastructures in diverse contexts. Institutions seeking to integrate law, design, and technology through community-led experimentation can begin by creating spaces for transdisciplinary collaboration, investing in dual literacy training, and co-developing participatory governance mechanisms with affected communities. Whether embedded in law schools, civic tech hubs, design studios, or community organizations, the model can be adapted to local needs and capacities. It invites educators, legal practitioners, policymakers, technologists, and organizers to reimagine their roles, not as distant experts, but as co-learners and facilitators of civic agency. By adopting and evolving this approach, institutions can contribute to building ecosystems where legal empowerment, ethical AI governance, and collective care are not abstract ideals but lived and shared practices.

References

Alvesson, M., & Sandberg, J. (2011). Generating research questions through problematization. *Academy of Management Review*, 36(2), 247–271. https://doi.org/10.5465/amr.2009.0188

American Bar Association. (2023). Rule 5.5: Unauthorized practice of law; multijurisdictional practice of law. In *Model rules of professional conduct*. www.americanbar.org/groups/professional_responsibility/publications/model_rules_of_professional_conduct/rule_5_5_unauthorized_practice_of_law_multijurisdictional_practice_of_law/

Artificial Intelligence Index Report. (2024). Artificial Intelligence Index Report 2024. Human-Centered AI Institute, Stanford University. https://aiindex.stanford.edu/report/

Blackburn, L. (2023). Qualifying work experience: Do street law projects provide a "legal service"? *The Law Teacher*, 57(1), 84–88. https://doi.org/10.1080/03069400.2022.2153311

Chien, C. V., & Kim, M. (2025). Generative AI and legal aid: Results from a field study and 100 use cases to bridge the access to justice gap. *Loyola of Los Angeles Law Review*, 57(4), 903–969. https://digitalcommons.lmu.edu/llr/vol57/iss4/2

Davies, T., Colom, A., Velkova, L., Poblet, M., Nobbs, L., & Kuneva, L. (2025, February 26). *Participatory AI: Forging shared frameworks for action*. Tech Policy Press. www.techpolicy.press/participatory-ai-forging-shared-frameworks-for-action/

Drumbl, M. L. (2023). Public education: Engaging with secondary education in schools. In A. Lawton, A. Morgan, D. Massey, & D. Castelyn (Eds.), *The International handbook on clinical tax education* (pp. 100–110). University of London Press.

Giles, C., & Ang, Y. (2025). What qualifies a law degree: surveying elective module provision in undergraduate legal education across England, Wales and Northern Ireland in 2023–2024. *The Law Teacher*, 59(1), 2–21. https://doi-org.ezproxy.brunel.ac.uk/10.1080/03069400.2024.2433327

Gonzatto, R. F., & van Amstel, F. M. C. (2022). User oppression in human-computer interaction: A dialectical-existential perspective. *Aslib Journal of Information Management*, 74(5), 763–764.

Harbers, M., & Overdiek, A. (2022). Towards a living lab for responsible applied AI. In D. Lockton, S. Lenzi, P. Hekkert, A. Oak, J. Sádaba, & P. Lloyd (Eds.), DRS2022: Bilbao (pp. 1–14). Design Research Society. https://doi.org/10.21606/drs.2022.422

Karpen, I. O., & Senova, M. (2021). Designing for trust: Role and benefits of human-centered design in the legal system. *International Journal for Court Administration*, 12(3), Article 7. https://doi.org/10.36745/ijca.422

Kidd, C. D., Abowd, G. D., Atkeson, C. G., Essa, I. A., MacIntyre, B., Mynatt, E., Starner, T. E., & Newstetter, W. (1999). The aware home: A living laboratory for ubiquitous computing research. In N. A. Streitz, J. Siegel, V. Hartkopf, & S. Konomi (Eds.), *Cooperative Buildings: Integrating Information*, organizations, and architecture (Lecture Notes in Computer Science, Vol. 1670, pp. 191–198). Springer. https://doi.org/10.1007/10705432_17

Leminen, S., & Westerlund, M. (2019). Living labs: From scattered initiatives to a global movement. *Creative and Innovation Management*, 28(2), 250–264. https://doi.org/10.1111/caim.12310

Lim, D. (2025). Determinants of socially responsible AI governance. *Duke Law & Technology Review*, 25, 184–232.

Pro Bono Week. (2021, October 29). Where it all began: Pro Bono Week 2001–2021. Pro Bono Week.

Puzio, A. (2025). The entangled human being – A new materialist approach to anthropology of technology. *AI and Ethics*, 5, 2348–2349. https://doi.org/10.1007/s43681-024-00537-z

Schuurman, D., De Los Ríos White, M. I., & Desole, M. (Eds.). (2025). Living lab origins, developments and future perspectives. European Network of Living Labs (ENoLL), 18–45. https://doi.org/10.5281/zenodo.14764597

Simkovic, M., & McIntyre, F. (2014). The economic value of a law degree. *The Journal of Legal Studies*, 43(2), 249–289.

Smith, R. F. (2025, January 20). *Digital and wealth gaps have no place in the intelligent age: Here's how everyone can benefit from AI*. World Economic Forum. www.weforum.org/stories/2025/01/digital-divide-intelligent-age-how-everyone-can-benefit-ai

Taylor, L., de Souza, S. P., Martin, A., & López Solano, J. (2025). Governing artificial intelligence means governing data: (Re)setting the agenda for data justice. *Dialogues on Digital Society*, 2, 1–18. https://doi.org/10.1177/29768640241306800

Tedeschi, M. (2024). Datafication and urban (in)justice: Towards a digital spatial justice. *Geography Compass*, 18, e12763, 1–13. https://doi.org/10.1111/gec3.12763

Section II

Engagement in Shaping Democracy

4 Democracy by Design

Applying Latin American Lessons to Boston's First Participatory Budgeting Process

Maria Ramirez Cifuentes, Estefania Ciliotta Chehade, Angelique C-Dina, and Michael Arnold Mages

Introduction

Understanding the distinct social and economic conditions that gave rise to Participatory Budgeting (PB) and Participatory Design (PD) can offer valuable insights for creating more inclusive and equitable forms of civic collaboration, power distribution, and material production. While we acknowledge that both share significant surface-level similarities in their efforts to promote equity, justice, and the distribution of agency, there are deeper differences in the types of human needs they address. The objective of this study is to apply Max-Neef's framework of fundamental human needs to analyze the distinct purposes and histories of Participatory Design and Participatory Budgeting and to document how this understanding informed more meaningful engagement, ultimately going beyond the simple implementation of a PB or PD process. We will also detail the role of technology in enabling and facilitating this process.

We explore debates surrounding participation within the context of participatory democratic practices in civic spaces, relating our experience designing the PB pilot project in Boston. We aim to show how PD—both as a philosophy and as a method—can contribute to more just and equitable democratic futures. We do this by drawing lessons from the kinds of needs that shape the deliberative model of participatory governance, as seen in PB.

Through this study, we seek to examine how the distinct fundamental human needs addressed by Participatory Design (PD) and Participatory Budgeting can inform the design of more equitable civic engagement processes. This leads us to several subordinate questions: What fundamental human needs do PD and PB address? How do these needs and contexts shape participation barriers and opportunities? How can design interventions bridge these different needs-based approaches?

We asked, how might we use design methods to further PB and PD foundations and foster more equitable and democratic engagement through community-driven, deliberative conversation workshops? To answer this question, we will first explain the foundational concepts of analysis, focusing on human needs as outlined by Max-Neef, and discuss how participatory processes, in general, address these needs. Next, we will explore the histories of PD and PB separately, examining the distinct socio-political contexts that gave rise to each, and discuss how they address different human needs. Then, we analyze how understanding these different origins has led to varying approaches to engagement and equity, drawing from our work with the City of Boston. Finally, we end this article with some reflections and lessons learned, hoping they will serve as practical guidance for future work in this space.

DOI: 10.4324/9781003667872-6

Max-Neef's Fundamental Human Needs and Participatory Practices in Latin America: Historical Overview

There are distinct social and economic conditions that gave rise to PB and PD, and the needs they address are a direct reflection of their respective contexts of origin. PD originated in Scandinavia, and it aimed to democratize the workplace, giving workers agency in shaping technological and organizational change. PB, on the other hand, held more ambitious goals—namely, the restructuring of state–citizen relations and the expansion of democratic participation in Brazil.

Understanding the historical and theoretical underpinnings of these approaches provides a necessary foundation for developing participatory strategies that are more responsive to contemporary struggles in the context of economic inequality, political disenfranchisement, and technocratic governance. These differences across regions expose how participatory methods can either reinforce existing institutional frameworks and social orders or challenge them in radical ways. While some approaches operate within established systems, others seek to transform them. To explore this further, we provide a historical overview of Latin America's long and complex history of participation.

Participation in Latin America is a hard-won achievement that can be traced back to the period of independence. This process unfolded differently across the region due to varying colonial experiences and regional conditions. Brazil experienced a gradual path to autonomy that lasted nearly 18 years, culminating in 1822, while Spanish America underwent a more rapid and widespread independence movement (Hamnett, 1997). This process of independence profoundly influenced and shaped identity and participation as we know them today.

While there are many differences in how the process unfolded across the region, certain factors are shared across most countries. For example, the leaders of the independence movements were often elite groups who sought to redistribute power. As the conflicts progressed, however, popular participation grew—sometimes emerging spontaneously, other times encouraged by movement leaders. This shift brought new social dynamics into play, involving more diverse people in matters of shared concern. However, the extent of popular involvement and social disruption varied greatly from country to country. In places like Venezuela, for example, the lower classes had a profound impact on the movement's outcomes, while in other states, like Perú, Brazil, and Mexico, the elites maintained strong control over the process (Hamnett, 1997).

Once independence was achieved, the interests of the elites were prioritized, and national identity and sentiment largely became constructs promoted by the newly formed states. Despite invoking the ideals of popular sovereignty, these elites were not interested in creating open political systems or functioning democracies (Hamnett, 1997). Power became concentrated in the hands of a few, paving the way for the rise of authoritarian regimes.

This independence movement also left a lasting mark on the region's political structure. The increasing militarization of society gave military leaders considerable political influence, often translating into political power in the post-independence era. This militarization is a key factor in the later emergence of dictatorships across the region, which also had a profound influence on participation.

In terms of governance, in most countries, national oligarchies controlled state development, creating "liberal democracies" that, in practice, excluded the majority (Max-Neef et al., 1991). This exclusion gave rise to populist regimes that sought to increase participation and modernize society, but they did so through universal suffrage and sectoral mechanisms, offering limited opportunities for direct citizen involvement. For the most part, states assumed a regulatory role, implementing redistributive policies.

This form of governance was driven by "homogeneous national projects" that limited genuine social participation through authoritarian or paternalistic models. Many newly independent Latin American states adopted European models of nationhood that prioritized cultural and linguistic uniformity. This often led to efforts to impose a singular national culture, marginalizing or suppressing indigenous and minority cultures in the process. The elites' embrace of nationalism was aimed at shaping post-colonial societies in their own image, further excluding the majority of the population from the nation-building process. This notion of homogenization created a context in which the diverse nature of civil society was undermined, and a wide range of voices and needs were suppressed. Social movements, therefore, had to fight for inclusion in the discussion of policies that affected them. Examples include the Zapatista movement in Mexico (Avila, 2018), the landless workers' movement (MST) (Rocha & Branford, 2002) and Workers Party (PT) (Lowenthal & Keck, 1992) in Brazil, and the indigenous rights movements in Bolivia (Albro, 2006) and Ecuador (Selverston, 1997). Historically marginalized groups—those most oppressed and disadvantaged—had to struggle hardest for recognition, and many of these systemic issues remain unresolved today.

Building on this historical context, Manfred Max-Neef, a Chilean economist, argued that understanding human needs could bridge the gap between philosophical anthropology and political action. Drawing from his experience living with and researching the poor across Latin America, he sought to develop a theory of human needs that was both understandable and practical for development. In essence, he proposed that a theory of human needs must take us from the social science perspective "humans are like this" to a proactive, designerly perspective that "society should be organized in this way" (Max-Neef et al., 1991).

This perspective frames needs as motivators and resources that drive human action. Needs reveal the essence of a person in a compelling way, showing both their limitations (deficiencies) and their potentials (capabilities). However, Max-Neef challenges the traditional conception of needs, setting his work apart and making it particularly relevant to this discussion.

Max-Neef argued that the common belief that human needs are infinite, constantly changing, and culturally specific is not entirely accurate. Instead, he proposed that human needs are finite, universal, and consistent across cultures and historical periods. What differs across cultures and time periods is the way those needs are satisfied. Different economic, social, and political systems adopt various approaches to meet the same fundamental human needs. In each system, needs are addressed (or neglected) through the generation (or lack) of different types of "satisfiers" which are not just material economic resources but also all the ways of being, having, doing, and interacting in the environment that contribute to the fulfillment of these human needs. Therefore, determining the degree to which individuals can influence the structures that shape their opportunities is crucial.

All of this is represented in a framework of needs and satisfiers included in Max-Neef's work. The framework is a matrix with axiological categories (nine fundamental needs: *subsistence, protection, affection, understanding, participation, leisure, creation, identity, and freedom*) on one dimension and existential categories (four modes of expression: *being, having, doing, and interacting*) on the other. Within this matrix, Max-Neef suggests that there are paradoxical situations where actions intended to fulfill a need may instead destroy the possibility of fulfilling it and negatively impact other needs. Out of the 9 needs, the five in Table 4.1 are especially prone to these paradoxes and more salient to PB.

In the upcoming section, we draw out the contradictions to connect to our central argument about how PB and PD respond to different human needs. We also examine how participatory initiatives emerging from these paradoxical contexts hold the potential to challenge dominant

Table 4.1 Needs and satisfiers

Fundamental human needs	*Being (qualities)*	*Having (things)*	*Doing (actions)*	*Interacting (settings)*
Subsistence	Physical and mental health, equilibrium, adaptability	Food, shelter work	Feed, procreate, clothe, rest, work	Living environment, social setting
Protection	Care, adaptability autonomy, equilibrium, solidarity	Insurance systems, savings, social security, health systems, rights, family, work	Cooperate, plan, take care of, cure, help	Living space, social environment, dwelling
Participation	Receptiveness, dedication, solidarity, determination, respect, passion, sense of humor	Responsibilities, duties, work, rights	Become affiliated, cooperate, propose, share, interact, dissent, express opinions	Associations, parties, churches, neighborhoods, family
Identity	Sense of belonging, self-esteem, consistency, assertiveness, differentiation	Symbiosis, language, religions, work, habits, customs, values, norms, sexuality, historical memory, work	Get to know oneself, grow, commit oneself, integrate oneself, decide on, get to know oneself, actualize oneself	Social rhythms, places one belongs to, everyday settings, maturation stages
Freedom	Autonomy, passion, self-esteem, open-mindedness	Equal rights	Dissent, choose, run risks, develop awareness	anywhere

Source: Manfred Max-Neef.

models of oppression that have historically undermined societal well-being. These efforts offer valuable insights for strengthening civic engagement.

One of the key differences between PD and PB becomes clear in this framing. Participatory design originated in societies with a history of progressive social policies, a shared cultural background, and relatively stable political systems. Emerging in Scandinavia in the late 1960s, it was driven by worries about automation, surveillance, and information technologies diminishing workers' influence in decision-making. Participatory design sought to counter this by involving workers in the design of these new technologies. It is a practice committed to democracy, labor, and equity at its core (Simonsen & Robertson, 2013).

PB, in contrast, is a practice that emerged as an antipoverty measure by the collective action of the Workers' Party in 1989 post-dictatorship Brazil to allow citizens to directly decide how a segment of public funds is spent (Bartocci et al., 2023).

It involves stages such as idea generation, evaluation, voting, and project execution, with both immediate, tangible community improvements and long-term benefits like increased civic involvement and trust in government. It's a deeply experiential practice, with people playing an active role in nearly every stage. It is a systematic response to the paradoxical legacies of post-dictatorship where citizens generate and deliberate on proposals to address community needs (inputs), a designated organism within the larger government implements them (outcome), and the government as a whole is held accountable for its delivery, as Figure 4.1 shows.

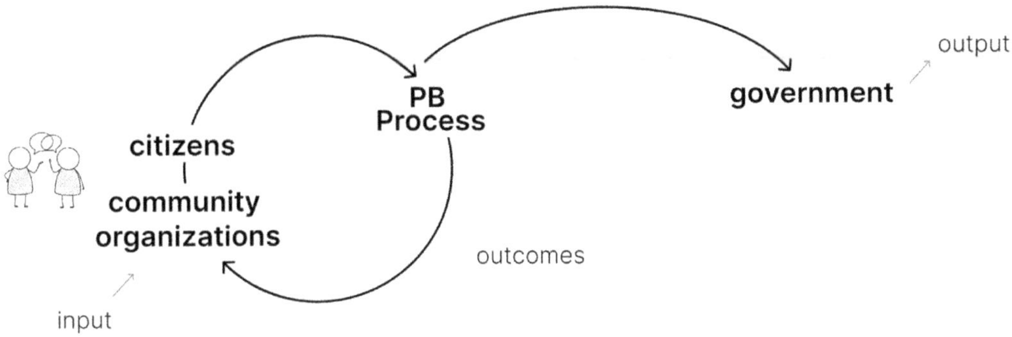

Figure 4.1 Participatory budgeting process feedback loop.

Source: Authors.

Participatory Processes Across Different Socio-political Contexts: The Impact of The Paradoxes

Dictatorships in Latin America often claimed to protect national interests and social order, but their methods resulted in significant deprivation of other fundamental needs (Max-Neef et al., 1991; see Table 4.2).

Participatory Budgeting challenged that dynamic by creating a new space where poor and historically excluded communities could have a real say in public decisions, challenging these paradoxes both for their individual and collective benefit. The inclusive features of PB assumed a particularly transformative and democratizing significance in a continent marked by social inequalities and elite dominance in participatory governance from the late 1980s onward. These participatory practices not only gained legitimacy but also became institutionalized, resulting in widespread civic engagement in the formulation of public policy. (Lüchmann et al., 2018)

In Argentina, for example, PB was implemented at the municipal level and later supported by national efforts through the National Program of Participatory Budgeting (PNPP), which operated between 2007 and 2015. This program aimed to strengthen and expand mechanisms for citizen involvement in governance (Arena, 2018). In Colombia, the first PB initiatives took place around 1991 with the adoption of a new constitution that emphasized human rights, participation, and decentralization. The Porto Alegre model was used as a reference for local adaptation across the country (Jaramillo, 2018). Perú took a step further in 2002, passing a national law that made PB mandatory in all regional and local governments (McNulty, 2018).

These examples demonstrate how participatory practices that emerge in response to significant deprivation of needs have the potential to influence the structural conditions that shape political inclusion, civic empowerment, and engagement over time (Lüchmann et al., 2018). Understanding this helps resist the tendency to transfer participatory methods from one context to another without depth, resulting in processes that invite users in name but not in influence. Genuine transformation depends not just on the form of participation, but on the depth of its intent and the structures that support it. Connecting this historic understanding with a genuine commitment to the political and ethical principles of PD is key to creating spaces where participation isn't just symbolic but actually reshapes how decisions are made while fostering real, meaningful engagement.

Table 4.2 Paradoxes and participatory budgeting response

Paradox	*Description*	*Participatory budgeting response*
Subsistence Paradox	Economic policies designed to safeguard national interests frequently led to extreme inequality and increased poverty	Empowers citizens to directly influence how resources are allocated, fostering a sense of security through community-driven decision-making and reducing inequality through collective input
Participation Paradox	The will of the people, masked as representation through a single leader, dismantled participatory structures	Reinstates meaningful participation through deliberative process where citizens are not only informed but also understand and commit to civic life
Identity Paradox	National projects that sought to create a unified identity marginalized the diverse cultural, ethnic, and social identities of minority groups	Validates pluralistic identities (Escobar, 2018) that had previously been suppressed
Freedom Paradox	Expressions of opposition resulted in severe consequences, limiting freedom	Creates safe space for open dialogue and community expression

Note. The left column (paradoxes) adapted from Manfred Max-Neef; the right column (PB responses) created by the authors.

In the following section, we will detail the practical application of these concepts through our work with the City of Boston.

Participatory Budgeting in Boston: Collaboration, Framework, Toolkit Design, and Impact

In 2024, the authors partnered with the Office of Participatory Budgeting (OPB) in Boston, Massachusetts. This effort was led primarily by the OPB and the Northeastern University Center for Design (CfD), and it represented a collaborative approach to co-designing for participation as an attempt to increase community and citizen engagement during the initial stage of the PB process: idea generation. Broadly, the PB process consists of multiple cycles carried out within a one-year span. Figure 4.2 provides a more detailed overview of this process.

The strategic design team—OPB representatives and CfD researchers—developed frameworks and tools to enhance community participation through cross-cultural conversation design and systems innovation. The approach to designing for participation included fundamental strategies for its success: (1) the delegation of community engagement to mid-level organizations, (2) carefully considering the design and context for deliberation, and (3) the incorporation of technology used to help marginalized community members participate, share their needs, and foster a greater sense of shared ownership over the process.

This case study draws on data collected during our role as the strategic design team, working in collaboration with the External Oversight Board. Working within a design consultancy model, we developed materials and facilitation agendas for resident workshops, while the workshops themselves were conducted by community partners under contract with the city. This structure ensured that all community engagement was led and mediated by trusted local organizations rather than academic researchers or government employees. All participant data referenced in this analysis was collected by the City of Boston through the official PB website,

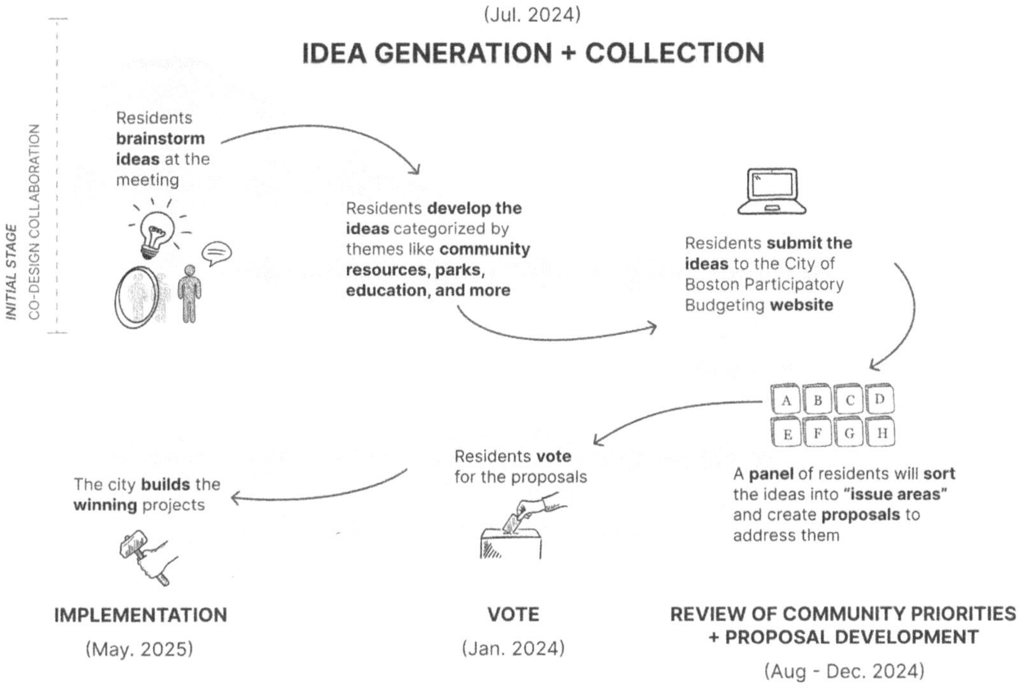

Figure 4.2 Participatory budgeting process in Boston.

Source: Authors.

maintaining appropriate institutional oversight and data governance protocols in compliance with Massachusetts open data laws throughout the process. Additionally, all evaluation activities were conducted by a third-party vendor hired by the city, ensuring independent assessment. We describe our approach and the collaborative work in the next section.

Stakeholders, Relationships, and Collaboration

The OPB was keen on building relationships with multiple stakeholders at different levels across the city as pillars of participatory engagement. They include the External Oversight Board, diverse city departments, residents, community partners, and community organizations.

The External Oversight Board was composed of nine experienced members in different areas such as finance, urban planning, and community organizing, among others. They included residents, community members, non-profit organization leaders, and retired municipal officials. Their role was to monitor the process and provide feedback every step of the way, making sure the community's needs were always prioritized.

To start the process of connecting with community organizations and other members, the OPB held a community meeting at a local design collaborative, the Design Studio for Social Intervention (ds4si).[1] They invited different community organizations to brainstorm and shed light on potential challenges for PB implementation and engagement. This first meeting helped us understand the lack of trust that communities, especially marginalized and neglected ones, have in the city government and its initiatives. It helped distill insights on how to go about building relationships with communities the city serves, as well as design approaches to address communities' needs and desires more critically and mindfully, while thinking about avoiding the paradoxes elicited earlier by Max-Neef.

The authors met weekly with the OPB team to align goals and deliverables and review the work progress. Our meetings were hosted mostly online, with three in-person collaborative sessions.

The team also had iterative engagements with community partners and the PB External Oversight Board. Community organizations had to apply to partner with the OPB, and they were fundamental in engaging with both city officials and the strategic team on one hand and members of their communities on the other to invite them to participate and engage with the Idea Generation Workshops. This network of community partners included organizations that operate in historically underserved neighborhoods such as *Sociedad Latina, Mass Senior Action Council, Maverick Landing/NUBE, Center for Teen Empowerment, Boston Center for Independent Living, Allston Brighton Health Collaborative, Transformational Prison Project, Union Capital Boston, and VietAID.*

In addition, two other organizations partnered with the OPB and the CfD to assist with two specific tasks: PoePublic[2] developed the idea collection website, and Data+Soul[3] surveyed and evaluated this first cycle.

Fostering and developing this network of conversations within the city allowed us to leverage a diversity of perspectives that enabled us to think systemically and systematically in designing the Idea Generation Workshop Toolkit. Supporting citizens' participation and engagement was a pillar throughout our conversations and planning meetings. Eliciting and facilitating conversations and the exchange of citizens' needs were incorporated as a fundamental layer in our toolkit and workshop design.

Idea Generation Workshop Design

To design and develop these workshops, we utilized aspects of different theoretical frameworks together—from design for conversations (Arnold Mages, 2024) and conversation as action (Dubberly & Pangaro, 2019), deliberative democracy (Gastil & Black, 2007), citizen empowerment through access to information (Dewey, 1988), public sphere (Habermas, 1985), and civic discursive spaces as intermediary space (Cornwall & Coelho, 2007).

In designing the workshop, we aimed to create space for dialogue around community values and priorities, moving away from deficit thinking. Through several iterations, we reframed "needs" as "priorities"—a shift reflected in our final toolkit.

These workshops were envisioned to create welcoming and inclusive spaces, physical and digital, to foster conversations between city representatives, partner organizations, and their constituents, and allow them to share ideas and participate in the decision-making process.

Each workshop was led by a facilitator chosen by the community partner organization and adapted to fit local needs. The goal, whether online or in person, remained the same: to help residents express concerns and, through conversation and creative activities, shape them into actionable ideas for the city.

The Space

The workshops were designed with a focus on supporting participants' well-being within the meeting space. Our participants would come to PB forums from work, school, family matters, and other challenging commitments. Some participants needed childcare. All of our participants would eventually need food for fuel and replenishment. To that end, all the organizations contracted as forum hosts were encouraged to spend part of the event budget on food and drinks and provide childcare or, at a minimum, a child-friendly atmosphere. This welcoming environment signaled that the city and partner organizations were actively working to facilitate participation.

Instead of following a traditional, linear educational format, the workshops were crafted as interactive and engaging experiences with tangible materials. This allowed all participants, regardless of their familiarity with budgeting concepts, to contribute equally in a welcoming space.

The Toolkit and the Idea Collection Website

We designed a flexible and adaptable toolkit after several iterations incorporating feedback from the different stakeholder groups. The toolkit was designed as an engagement tool. It included a moderator's and participant's agenda, a briefing document, a project builder (Card Game) and a worksheet, guidelines for participation, and a hosting a meeting checklist. In use, it became a symbol of the city's commitment to PB with a special focus on inclusivity and fairness in the approach (Figure 4.3).

The Briefing Document was designed to prepare residents for the workshop; it provided an overview of the process and the goal of the activities (see Figure 4.4), and it served as a roadmap, sharing foundational PB knowledge and expectations for each activity.

The Briefing Document was accompanied by a slide presentation that provided visual representations of the information and roadmap. Both reflected the overall goals above.

The participants' and moderators' agendas were designed to guide participation and provide a clear progression of events. In the case of the Participant's Agenda, it served to set the stage and prepare participants for each activity throughout the workshop. It also served as a "cheat sheet" for participants to refer back to and check the goals and tasks on each activity. The moderator's agenda, on the other hand, provided a more comprehensive guide about the progression of events. Because these were designed to be adaptable to fit each partner organization's needs and constituents, the moderator's agenda had a more in-depth description about each activity, the goal for each activity, prompts and questions as suggestions for promoting more engaging and inclusive conversations, as well as the suggested duration for each activity. Providing the goals, description, and suggested prompts and questions per activity was designed to communicate the underlying logic of the design choices, offering more agency and autonomy for each moderator/facilitator to adapt these as they saw fit. Each activity, and the prompts and questions incorporated, was carefully designed taking into account our theoretical frameworks described in earlier sections, while promoting a safe space for residents to share their concerns and needs and translating those into more hopeful and actionable projects and ideas (Figure 4.5).

The PB Project Builder (card game) served a dual purpose: to stimulate innovative thinking and to familiarize participants with accepted project categories and their feasibility. It provided an opening, a view into the city structures, revealing the city's capacity to respond to citizens' needs. It was designed to assist in the "reframing" of needs and, in a playful manner, allow participants to transition from the "deficit state" into more hopeful and creative thinking.

We designed three different types of cards. The category cards helped define the departments or areas within the city a project/idea could apply to; the place cards represented the potential scale of focus of ideas/projects—whether the impact was more at a street level, neighborhood level, or had the potential to impact the city as a whole; and finally, the type cards helped explore

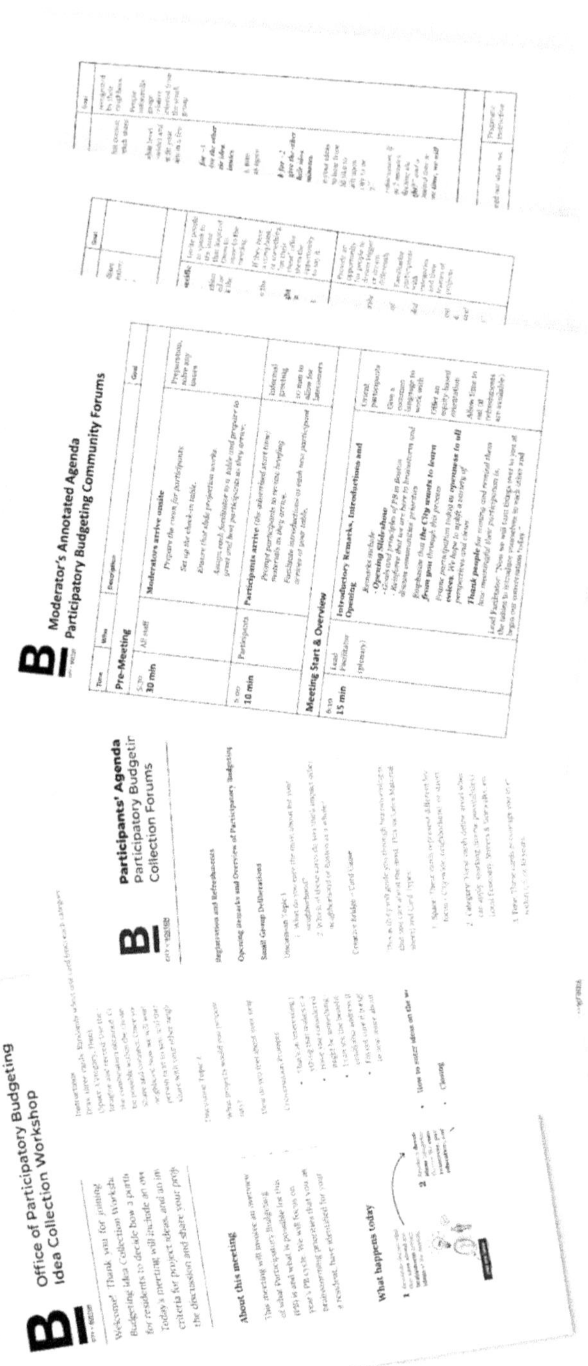

Figure 4.3 Toolkit overview: briefing document and workshop agenda (moderator and participant).

Figure 4.4 Workshop overview—illustration by authors included in the briefing document.

Figure 4.5 Toolkit in action: Idea Generation Workshops Summer 24'.

different proposal options as to whether they fit under infrastructure projects or as programs and services.

The PB Project Builder was accompanied by a worksheet to allow participants to quickly write down ideas for projects, identifying their category and impacts. The worksheet was also envisioned as a placeholder for participants to expand on their ideas and submit them later on the city's website for idea collection (Figure 4.6).

Figure 4.6 PB Project Builder.

The guidelines for participation mainly provided a set of suggestions for both facilitators and participants on how to foster a welcoming and safe environment for all, respecting diversity of perspectives, promoting engagement and participation, being an active listener, sharing and building on ideas, and creating the collective and collaborative space for ideas to emerge.

Finally, the hosting a meeting checklist was designed for partner organizations to ensure they had all necessary resources for a successful Idea Generation Workshop. It included community outreach strategies, logistics and setup, planning for translation or interpreters, childcare, and food, among others.

In addition to the toolkit, a website[4] was designed by a partner organization, PoePublic. The website served as a collection tool and was envisioned to be the endpoint for all resident ideas.

Reporting: Outcomes and Impact

For the first iteration of these Idea Generation Workshops, which took place in July 2024, the OPB collaborated with 19 Community Partners tasked with organizing and gathering constituents to participate in the workshops: 14 were hosted in person and 5 online. Five hundred and sixty residents participated in these workshops, with 312 ideas submitted (Data+Soul, 2024). At the end of each workshop, the city—together with Data+Soul, the expert evaluation agency—handed out surveys to participants and held debrief sessions with both residents and community partners. The evaluation focused on four key areas:

- **Participation**: how well were priority populations represented
- **Impact**: changes in trust in local government, civic knowledge, and interest in future engagement
- **Process**: what helped or blocked participation
- **Equity**: who was reached, and how fairly

Data Sources: The portal for idea submission collected neighborhood information from all participants. Demographic details such as race, ethnicity, income, and age were collected voluntarily. The workshops engaged a higher proportion of Latinx and Black residents compared to their share of Boston's overall population. Participants with lower incomes were especially active in the process.

In total, this first cycle allowed the city to reach 789 residents contributing with ideas and participating in the process. Every neighborhood, especially underserved ones, engaged with the process. One thousand two hundred and thirty-eight ideas were submitted to the website. (Data+Soul, 2024)

Overall, participants enjoyed the workshops, especially discussing and building community around local issues. Eighty-six percent of workshop participants who completed the exit survey indicated they felt their voices were heard by the city. A good sign of growing trust in the city. Partners valued the support, training, and materials (incl. translations) provided by OPB. These early results suggest the city's model—working with local partners and mid-sized organizations—was effective in getting more people involved. (Data+Soul, 2024)

Technology also played a key role, from online forums to the submission platform. Making the process public and accessible helped redistribute agency within the Participatory Budgeting framework, increasing transparency, accountability, and inclusivity. It also provided the OPB with the flexibility and speed needed to scale up the process, reaching a broader audience (Table 4.3).

Table 4.3 Ideas in action evaluation categories and details of findings

Category	*Details*
Organizers	OPB + 19 Community Partners
Workshop format	14 in-person, 5 online
Participation	789 residents, 1238 ideas submitted, 13 neighborhoods reached
Data collection	Neighborhood info (all participants); demographic data (optional:race, ethnicity, income, age)
Key demographics engaged (Equity Outcome)	Latinx and Black residents; lower income participants; Youth and older participants; (most active)
Projects selected (equity outcome)	All 6 projects selected addressed needs of priority populations
Participants impact	86% of participants felt heard by the city
Partners feedback	Valued the support and materials provided by the City

How the Office of Participatory Budgeting Fostered Engagement and Focused on Online Strategies

On the residents' front, the OPB employed various strategies to foster engagement. Some of them were intended to reach out to the public in general, and some were more targeted. For example, the OPB, in collaboration with the Oversight Board, hosted a series of online meetings to update the public on the development of the PB initiative. These meetings were open to anyone, creating a snowball effect that helped engage a broader population.

Also, for broader engagement, the OPB used more traditional outreach channels. Banners with QR codes were placed throughout the city and on public transportation, directing people to the initiative's website.

In terms of the targeted strategies, the partnership model to host the workshops with communities from marginalized groups was key in reaching more people. Reports from the evaluation agency highlighted that participants in the workshops often recommended others to join future events hosted by the same community partners. This was a successful tactic because the community partners didn't stop at just one workshop; they held between two and four sessions to maintain momentum and reach those who might not have heard about the events or were unable to attend for various reasons. This flexibility also allowed participants to fit the workshops into their already complex schedules.

During the voting phase, the OPB expanded its outreach strategy through social media. They collaborated with the Mayor's Office to create social media content shared on the city's official accounts, with videos of the mayor herself encouraging residents to participate. The OPB also worked with the Office of Neighborhood Services to promote the initiative in their monthly newsletters, while other departments shared content via their own channels. Media outlets, including newspapers and ethnic radio stations, were also involved in spreading the project.

While technology played a role in giving more visibility to the project, one of the most notable ways technology contributed to this process was by ensuring accountability and keeping the public informed. The website became a sort of living document where submissions were openly visible to all, promoting transparency and engagement. Also, in line with the project's core values, the website was designed with technical features that reflect Boston's diverse population, offering multiple languages so people from various backgrounds could participate and stay updated on the process.

Lessons Learned

While PB holds the promise of fostering genuine engagement, promoting inclusion, enhancing government transparency, and addressing inequalities, its implementation in the North American context presents distinct challenges compared to its application in the Global South. Established infrastructure and more secure access to basic needs in a much wealthier society may lead to political apathy and diminish motivation for public participation. This apathy, often rooted in the historical neglect and/or exploitation of marginalized communities, was a determining factor in how we approached this project.

By reframing how these communities were brought into the process and their experience interacting with it, our approach helped bridge this gap. We built an infrastructure designed to empower people to actively be part of the process and shape the outcome, increasing their emotional and intellectual investment.

By incorporating these thoughtful elements—flexible scheduling, food, childcare, interactive engagement, and language accessibility—the workshops provided a model for inclusive PB. This approach emphasized that civic engagement should not be a privilege but a right accessible to all community members.

Our process enabled us to learn how PD can help uplift diverse community voices in Boston and aid in identifying and addressing barriers that might otherwise go unnoticed or neglected. We learned we need to often ask ourselves, "Who are we leaving out? Who does this initiative benefit?"

We understand that many underrepresented voices were not included, and we continue to advocate for the inclusion of those voices and for radical changes that align with the wishes of those communities. For instance, knowing that language has historically been a barrier to participation, we ensured the toolkit materials were also available in seven languages to welcome community members' participation. Languages such as Haitian Creole or Mandarin, other prominent languages in our community, were prepared as digital files, but printed versions were not requested by our community partners. Moving forward, we aim to ensure a broader range of language speakers use the toolkit.

By fostering engagement, reaching a broader audience, and facilitating more inclusive participation in our collaborative efforts, we aim to spark ongoing questions and inquiries.

Inclusive participation is a central tenet of PD. Implementing inclusive participation often involves using various methods to accommodate the needs of diverse groups. In our toolkit design, we ensured our facilitators had visual aids, workshop preparations, and portable materials for the meetings to enhance accessibility. Technology provided us with a path to facilitating engagement through the various online forums and platforms for idea submission (website), reaching a broader audience, and increasing accessibility, inclusivity, accountability, and transparency. This approach helped to level the playing field, creating a more democratic framework for engagement.

Conclusion

This work explored how aligning with the foundational goals of PB, which seeks to address fundamental human needs through inclusive, community-driven decision-making, can foster more just and equitable democratic futures.

Our design approach created workshops and toolkits that tackled participation barriers (language, childcare, access) and aimed to foster dynamic engagement and equitable outcomes.

Communities are knowledgeable about their needs for government services; design simply provides the accessible dialogue space they require.

Looking ahead, we hope the City of Boston continues to utilize PB and PD strategies to foster trust and relationships with community members while committing to actions and a process of community accountability. These approaches are essential for building civic trust and advancing social equity in urban planning and governance.

Notes

1 See Design Studio for Social Intervention, (n.d.), Home, www.ds4si.org/
2 See PoePublic (n.d.), Home, https://about.mjumbepoe.com/
3 See DataSoul (n.d), Home, www.dataplussoul.com/
4 See City of Boston Participatory Budgeting (n.d.), Home, https://participate.boston.gov/page/about

References

Albro, R. (2006). The Culture of Democracy and Bolivia's Indigenous Movements. *Critique of Anthropology*, 26(4), 387–410. https://doi.org/10.1177/0308275X06070122

Arena, E. (2018). Participatory Budgeting in Argentina (2002–2018). Advances and Setbacks in the Construction of a Participatory Agenda. In N. Dias (Ed.). *Hope for Democracy: 30 Years of Participatory Budgeting Worldwide* (pp. 123–133). Epopeia Records.

Arnold Mages, M. (2024). *Conversational Design: Improving Participation and Decision-Making in Public Organizations.* BIS PUBLISHERS B V.

Avila, O. (2018). La democracia zapatista. Entre la dinámica comunitaria y los mecanismos de regulación representativa. *Revista de Ciencias Sociales*, 31(42), 199–205. https://doi.org/10.26489/rvs.v31i42.9

Bartocci, L., Grossi, G., Mauro, S. G., & Ebdon, C. (2023). The Journey of Participatory Budgeting: A Systematic Literature Review and Future Research Directions. *International Review of Administrative Sciences*, 89(3), 757–774. https://doi.org/10.1177/00208523221078938

City of Boston Participatory Budgeting. (n.d.). https://participate.boston.gov/page/about

Cornwall, A., & Coelho, V. S. P. (2007). *Spaces for Change? The Politics of Participation in* New Democratic Arenas. Zed Books.

Data+Soul. (2024). *Participatory Budgeting Evaluation 2024–25*. www.boston.gov/sites/default/files/file/2025/03/OPB%20Final%20Presentation%20to%20EOB%203.20.25.pdf

Design Studio for Social Intervention. (n.d.). www.ds4si.org/.

Dewey, J. (1988). Challenge to Liberal Thought. In J. A. Boydston (Ed.), *The Later Works, 1925–1953: 1925*. Southern Illinois University Press. https://books.google.com/books?id=nu_B91DQiRcC

Dubberly, H., & Pangaro, P. (2019). Cybernetics and Design: Conversations for Action. In T. Fischer & C. M. Herr (Eds.), *Design Cybernetics* (pp. 85–99). Springer International Publishing. https://doi.org/10.1007/978-3-030-18557-2_4

Escobar, A. (2018). *Designs for the Pluriverse: Radical Interdependence, Autonomy, and the Making of Worlds*. Duke University Press. https://doi.org/10.1215/9780822371816

Gastil, J., & Black, L. W. (2007). Public Deliberation as the Organizing Principle of Political Communication Research. *Journal of Deliberative Democracy*, 4(1). https://doi.org/10.16997/jdd.59

Habermas, J. (1985). *The Theory of Communicative Action: Volume 1: Reason and the Rationalization of Society*. Beacon Press. https://books.google.com/books?id=kuFhjNZuHTAC

Hamnett, B. R. (1997). Process and Pattern: A Re-examination of the Ibero-American Independence Movements, 1808–1826. *Journal of Latin American Studies*, 29(2), 279–328. https://doi.org/10.1017/S0022216X97004719

Jaramillo, R. (2018). Participatory Budgeting in Colombia. In N. Dias (Ed.), *Hope for Democracy: 30 Years of Participatory Budgeting Worldwide* (pp. 135–145). Epopeia Records.

Lowenthal, A. F., & Keck, M. E. (1992). The Workers' Party and Democratization in Brazil. *Foreign Affairs*, 71(4), 205. https://doi.org/10.2307/20045348

Lüchmann, L., Romão, W., & Borba, J. (2018). 30 Years of Participatory Budgeting in Brazil: The Lessons Learned. In N. Dias (Ed.), *Hope for Democracy: 30 Years of Participatory Budgeting Worldwide*, (pp. 89–103). Epopeia Records.

Max-Neef, M. A., Elizalde, A., & Hopenhayn, M. (1991). *Human Scale Development: Conception, Application and Further Reflections*. The Apex Press.

McNulty, S. (2018). Mandating PB: Evaluating Fifteen Years of Peru's National Participatory Budgeting Law. In N. Dias (Ed.), *Hope for Democracy: 30 Years of Participatory Budgeting Worldwide* (pp. 147–159). Epopeia Records.

PoePublic. (n.d.). https://about.mjumbepoe.com/.

Rocha, J., & Branford, S. (2002). *Cutting the Wire*. PRACTICAL ACTION PUBLISHING. https://doi.org/10.3362/9781909013117

Selverston, M. (1997). The Politics of Identity Reconstruction: Indians and Democracy in Ecuador. In D. A. Chalmers, C. M. Vilas, K. Hite, S. B. Martin, K. Piester, & M. Segarra (Eds.), *The New Politics of Inequality in Latin America* (1st ed., pp. 170–191). Oxford University Press. https://doi.org/10.1093/0198781830.003.0007

Simonsen, J., & Robertson, T. (Eds.). (2013). *Routledge International Handbook of Participatory Design*. Routledge.

5 Engaging in the Safety Design of Sustainable Tourism Services

The SafetySpinner Tool for Participatory Design

Minna Virkkula, Hong Li, and Zhengya Gong

Introduction and Background

Designing for safety in tourism services requires building trust and promoting a positive user experience by understanding the diverse backgrounds, cultural contexts, and individual needs of customers. Even small details may influence the overall safety experience. This diversity poses challenges for service design, particularly in tourism, where expectations and environments vary widely.

This research aims to explore how participatory design can enhance tourism services by addressing the safety needs of users. Tackling these challenges requires innovative approaches that acknowledge the subjective nature of safety experiences and engage stakeholders, especially service providers, in co-creating sustainable solutions. This study focuses on the significance of safety experience design in services and explores how a new participatory design tool can enhance the design process, especially within adventure (Huddart & Scott, 2020; Taylor et al., 2013) and regenerative tourism contexts (Bellato et al., 2023; Luong et al., 2024). The objectives of this research are twofold: first, to examine the role of safety experience design in tourism services; and second, to develop and evaluate the SafetySpinner, a collaborative tool designed to support participatory safety design.

Safety risk in destination management is a well-researched topic. Safety in travel destinations encompasses managing risks such as crime, natural disasters, and health issues. This includes both objective measures (e.g., risk management and safety protocols) and subjective aspects (e.g., tourists' perceptions of safety) (Zou & Yu, 2022). To some extent, safety perception is subordinated to tourists' perception of the destination's safety image, which is essential for creating an ideal destination image (Chauhan, 2007; Zou & Yu, 2022). Additionally, several studies have examined risk factors and risk management in adventure tourism (Morgan, 2021; Dickson, 2012; Bentley et al., 2007). However, research on customers' safety experiences in services remains limited.

This study addresses the lack of customer-centric approaches in safety design by integrating participatory design principles with a deep understanding of individual safety perceptions. By acknowledging the subjective nature of safety experiences, this study focuses on the development of collaborative, user-centric safety design. The main research question is: How can participatory design methods be applied to address diverse user safety needs in tourism services through the development of a collaborative tool? The remainder of the chapter is structured as follows: a literature review on safety, demographics, and participatory design in tourism; a case study on the development and evaluation of the SafetySpinner; and finally, a discussion and concluding remarks.

DOI: 10.4324/9781003667872-7

Related Work

Participatory Design in Services

Participatory design is a methodology that actively engages users, stakeholders, and communities in the design process, ensuring that individuals affected by a design solution are involved in its development. Originating in 1970s Scandinavia within workplace democracy movements, it emphasized user empowerment and collective decision-making in technology development (Gregory, 2003; Simonsen & Robertson, 2012). Over time, participatory design has proliferated into domains such as Human-Computer Interaction (HCI), product design, urban planning, healthcare, and social innovation (Spinuzzi, 2005; Stephanidis et al., 2025). Its fundamental principles include democratic involvement, co-creation, iterative engagement, contextual understanding, and empowerment (Ehn, 2017; Muller & Druin, 2003).

Numerous factors influence participatory design, shaping the collaboration among users, stakeholders, and designers, consequently affecting the inclusiveness and outcomes of the design process. For example, understanding user demographics is crucial, as it allows designers to tailor solutions to diverse needs and preferences, thereby enhancing usability and relevance (Norman 2013). Recent research further emphasizes that integrating demographic diversity and accessibility considerations (e.g., for neurodiverse users or individuals with disabilities) is essential to achieving equitable and satisfying user experiences (Almenara et al., 2024; Gerling et al., 2024). This approach ensures that participatory design processes produce solutions that are inclusive and contextually meaningful for a wide range of stakeholders.

Safety and Democracy: Integral Components of Social Sustainability

Safety is a multidimensional concept, and the following key concepts are explained from the customer's perspective, focusing on various tourism services (Xie et al., 2021). Safety, as a component of social sustainability, is explored by Eizenberg and Jabareen (2017). The concept encompasses the right to safety and entails proactive measures to ensure protection and minimize risks, including improving infrastructure and safety protocols, as well as educating individuals on safety practices to prevent potential harm.

In the study by Bąk and Szczecińska (2020), several key demographic trends and their impacts on tourism, with a particular focus on safety, are examined. The aging global population is driving increased demand for accessible and safe travel options customized for older tourists (Cooper, 2022). Population growth varies by region, impacting tourist distribution and their service needs (Gidebo, 2021). In response to these trends, services should be tailored to diverse groups, particularly the elderly and solo travelers, by prioritizing personal safety. Ethical and environmental concerns also affect tourism and promote sustainable practices (Sharma & Sodani, 2023). Additionally, the shift toward interactive technologies enhances safety by enabling real-time support.

Customer Perceptions of Safety in Tourism

Safety encompasses multiple dimensions, and the following key concepts relevant to this study are explained from the customer's perspective in relation to tourism and its services. Safety can be examined from two perspectives based on destination characteristics and individual factors influencing travel decisions. Objective risk involves the measurable assessment of potential outcomes, while subjective safety is individually perceived, based on interpretations of possible outcomes (Karl & Schmude, 2017). From an objective perspective, safety and security in tourism represent a stable ecosystem that includes the management of risks and crises as well as the overall protection of tourists at destinations. This multifaceted concept involves customers' perceived sense of security, infrastructure performance, and destination management (Zou & Meng, 2020; Mansfeld & Pizam, 2006). The perceived safety of tourism destinations is based both on the individual tourist (personal traits, cultural background, socio-demographic factors) and on the destination itself, particularly how the actual risk situation is communicated or experienced (Karl & Schmude, 2017). Matiza and Slabbert (2021) identify three dimensions of subjective safety experience in tourism. Physical safety pertains to tangible aspects, such as the absence of physical hazards and preventive measures against harm, injuries, and accidents. Psychological safety relates to customers' emotional well-being, including their sense of security and peace of mind, which is influenced by factors such as stress, anxiety, and overall mental well-being. Social safety encompasses the social environment, shaped by social support, community cohesion, and interpersonal networks.

Narangajavana Kaosiri et al. (2019) provide insights into how different sources of user-generated content shape tourists' expectations and satisfaction. Their findings highlight that social media content contributes to destination image, risk perceptions, and visitor experience (Narangajavana et al., 2019). Moreover, social media serves as an effective tool for crisis communication in tourism (Barbe & Pennington-Gray, 2020). Conversely, from a subjective viewpoint, safety and security are psychological phenomena involving tourists' perceptions, awareness, concerns, and understanding of real events (Zou & Yu, 2022). Research shows that safety influences travel decisions and shapes the overall perception of tourism. These trends highlight the need for the industry to adapt to changing demographics and prioritize safety to meet evolving customer expectations. Understanding these perceptions is essential for effective management and long-term sustainability.

Preliminary Research and Methodology of the Case Study

This study builds on data collected from earlier projects to examine the participatory design of the safety experience in greater depth. It adopts research-through-design approach (Blessing & Chakrabarti, 2009), contributing to the development of foundational design knowledge and addressing safety considerations related to tourism and festival environments (Virkkula et al., 2024). The research incorporates the Adventure Tourism Risk Factors Model, which categorizes risks into external and operational factors. External risks include weather and terrain hazards, while operational risks involve customer skills, equipment, and management policies (Bentley et al., 2007). Customer safety concerns were further classified using a framework that divides them into physical, social, and psychological dimensions (Matiza & Slabbert, 2021). This comprehensive approach ensures that both objective and subjective aspects of safety are addressed.

To gather insights for a safety-focused tool and to identify common themes related to customer safety concerns in tourism services, we conducted a focus group with 22 participants ($F = 16$, $M = 7$). These participants were divided into five focus groups of four to six people to explore safety-related concerns and examine how key ideas interconnect. The participants were all third-year Finnish-speaking media and design students from the University of Lapland. Participation consent was obtained through a consent form explaining the purpose of the study and the use of its results. We utilized two methods in the workshop: card sorting and concept mapping. First, card sorting was used to identify common themes related to customer safety experiences, enabling participants to express how various themes relate to one another. In addition, this activity provided all participants with a shared understanding of customer safety concerns. Second, we used a concept map, which was applied to visually represent these connections, offering a clearer view of how different elements fit together within the safety experience.

The Card Sorting Activity

Each focus group received 35 safety aspect cards representing various aspects of customer safety concerns. The content for the initial set of cards was derived from earlier projects, which consisted of literature-based safety insights and involved co-design workshops with small tourism service providers in Northern Bothnia. Additionally, safety-related concerns were synthesized from previous research findings (Virkkula et al., 2024), providing a foundation for identifying recurring user-specific safety needs. A draft version of the SafetySpinner tool was used during the activity, including predefined categories for physical, psychological, and social safety, along with 19 associated cards based on Matiza and Slabbert's (2021) safety perception model. Participants worked in groups to sort the cards in their group into themes and added further concerns using blank cards. The participants also created titles for each theme, or subtitles when appropriate. Presented below are images from the card sorting activity (Figure 5.1).

Card sorting analysis was performed using similarity matrices. Thematic analysis was then applied to identify common themes, with cards grouped according to participants' choices. This method offered a structured approach to understanding how groups perceive and categorize information. We identified four common high-level themes from the card sorting activity: (1) environment and operational factors; (2) health and safety; (3) information, communication, and reliability; and (4) staff and interaction. Detailed insights for each theme are presented in Table 5.1. However, there are differences in the emphasis and approaches to various themes among the groups. For example, in the theme of environment and operational factors, groups emphasize different aspects such as weather conditions, the condition of facilities, or mobility. In terms of health and safety, the groups focus on different risk factors and safety perspectives. The groups for the sections on subjective safety were not altered or merged. Only a few cards were added to the physical safety section, specifically addressing crowding and inadequate equipment, as well as to the psychological safety section, particularly regarding employee

Figure 5.1 Images of card sorting activity conducted during the user study. Photograph by Minna Virkkula.

Table 5.1 Themes on safety factors based on card sorting

1. Perspectives on environment and operational safety		
Nature and environment	*Facilities (space and locations) and used equipment*	*Mobility*
Weather, seasons, encounters with animals, water and power outages	Lighting, condition of facilities, condition and availability of equipment, slippery surfaces, falling objects.	Route planning, traffic safety, congestion
2. Perspectives on health safety and risk of injury		
Medical emergencies	*Risks related to injury*	*Regulatory compliance*
Medical emergencies, health safety, panic situations	Risks related to activity, inadequate equipment, information reliability and sufficient.	Insurance, contracts
3. Perspectives on information communication and reliability		
Availability of information		*Cybersecurity*
Information reliability, up-to-date information, sufficient information before activities, independent of time and place		IT system disruptions, data security, financial transactions
4. Perspectives on staff and interaction		
Customer interaction		*Guidance*
Professionalism of employees, responding to feedback, interaction with customers		Guidance of the activity

behavior. Lastly, in the themes of information and communication, the groups highlight different aspects of information dissemination and the reliability of sources.

The Concept Maps

After the focus groups, we analyzed the concept maps according to their structure, content, and phrasing, along with conclusions on the essential elements in designing the customer's safety experience. Figure 5.2 illustrates the process of defining the concept map and the final presentation of a single map (G5).

Groups 1, 2, and 4 maps generated a diamond structure categorizing safety into psychological, social, and physical aspects. Figure 5.3 shows the concept map created by Group 2, which emphasizes the interconnectedness of these dimensions, highlighting how they overlap and influence one another. The map incorporates elements such as cultural and professional interactions, legal frameworks, health-related concerns, online safety, and environmental factors.

Our cluster analysis of the concept maps revealed key elements for designing a holistic customer safety experience in tourism services. All but one map highlighted the dimensions of social, physical, and psychological safety. Key recurring themes involved staff interaction, proper equipment, clear communication, and access to information. For example, phrases such

Figure 5.2 The process of defining concept maps and an example of the concept map (G5). Photograph by Minna Virkkula.

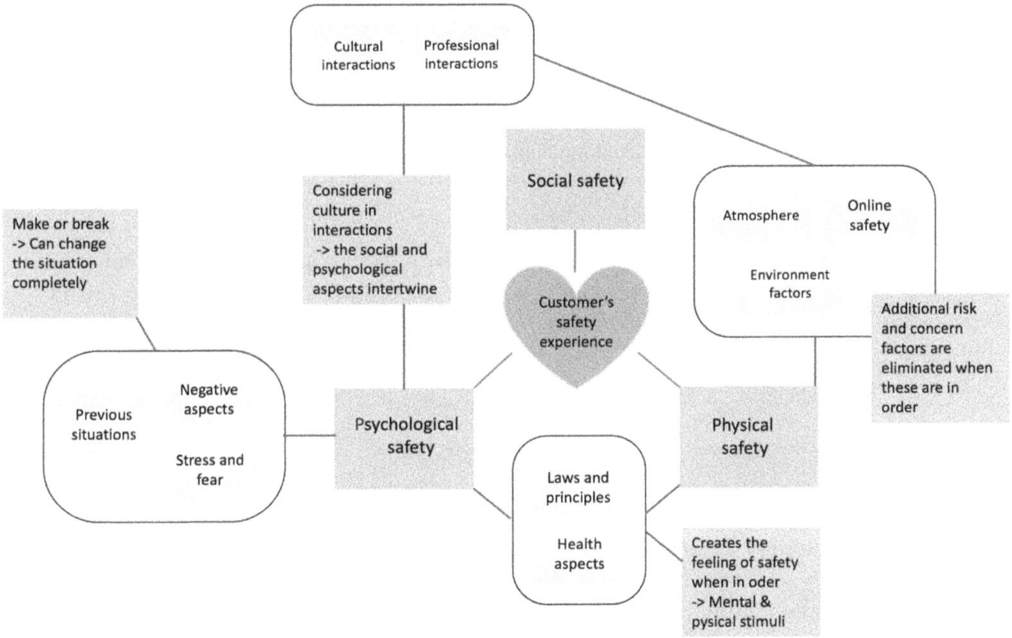

Figure 5.3 An example of the final concept map (G2).

Table 5.2 Concept maps overview

G1	Structure	The map forms a diamond structure, with social, physical, and psychological safety creating a triangular pattern.
	Content	Emphasizes staff interaction, facilities, equipment, contracts, and insurance and how they are connected to social, physical, and psychological safety.
	Phrasing	Staff responsibility, clear instructions, accessible information, and proper facilities and equipment are essential for service.
G2	Structure	The map generates a diamond diagram and divides safety into psychological, social and physical aspects.
	Content	A focus on physical, social, and psychological safety and how these intertwine with each other. Includes cultural and professional interaction, laws, health aspects, online safety, and environmental factors.
	Phrasing	Clarifying sentences of connections, noting that make-or-break moments can change situations due to past experiences, stress, and fear, and emphasize cultural interactions as part of professionalism.
G3	Structure	The structure of the map is a flowchart where different phases follow each other.
	Content	Background requirements must be met first, followed by risk management and an action plan. Trust will then develop.
	Phrasing	The map clarifies that safety is visualized in the map step by step, but there are many safety concerns that can fall into multiple phases. Timeliness, reliability and accessibility information must be ensured in all phases.
G4	Structure	The structuring themes are physical, social, and psychological safety and the map forms a diamond structure.
	Content	Content creates an interplay between physical, social, and psychological safety. Brought to attention internal and external risks, communication, and cyber security.
	Phrasing	Additional sentences and roles of operations for more detailed information. Emphasizes the importance of proper guidance and cultural awareness and outlines responsibility.
G5	Structure	The map was simpler in structure and shapes a network diagram related to physical, psychological, and social safety.
	Content	Highlights interaction in social safety, medical aspects, environment, equipment in physical safety and documentation and communication in psychological safety.
	Phrasing	No added information about relationships.

as "Suitable facilities and equipment for activity" and "Clear instructions prevent injuries" reflected these priorities. While some maps addressed cultural awareness and sustainability, others focused on legislation, health, online safety, or environmental concerns. The level of detail varied across groups, as shown in Table 5.2.

The maps reveal three key themes:

1. Subjective safety: Addressing customers' social, physical, and psychological safety in service design.
2. Practical focus: Emphasizing practical aspects of service provision:
    - ❖ Staff interaction: Ensuring that trained staff manage safety.
    - ❖ Availability of information: Providing timely and reliable safety-related information.
    - ❖ Facilities and equipment: Designing infrastructure to prevent harm.
3. A holistic safety approach: Addressing multiple dimensions of safety by considering physical factors, social interactions, and psychological comfort. This includes integrating cultural

considerations, ensuring cybersecurity and accessible information, and accounting for the operational environment, infrastructure, and mobility. These findings suggest the tool should adopt a holistic structure that incorporates diverse aspects of customer safety concerns and is adaptable to different tourism service contexts.

The Structure of the Participatory Design Tool for Safety

The design tool is structured based on core themes identified from card sorting and concept mapping. By structuring the tool around these interconnected themes, it prioritizes safety and supports participatory design in a service context. These themes reflect key elements of the customer safety experience.

Environment and operational factors: The physical environment (e.g., weather, season, nature, infrastructure, and mobility in operational environments) plays a crucial role in ensuring safety. Safe and well-designed operations in different environments, along with flexibility in adapting to changing conditions, reduce risks and contribute to a sense of safety.

Operations and collaborations of the stakeholders: Effective communication and collaboration within the company are essential for building trust and ensuring customer safety. Transparent and reliable communication can prevent misunderstandings and encourage collaboration. Facilities and proper equipment, along with thorough documentation and accessible information, are critical components of company operations. Continuous feedback is included for integrating all safety themes in order to identify gaps and areas for improvement.

Customers' subjective safety: Customers' social, psychological, and physical safety are closely linked. For example, interactions, atmosphere, customizability of the activity, and previous experiences can significantly impact well-being. Ensuring respectful and culturally aware interactions can further enhance subjective safety. As a customer, feeling safe involves a combination of these factors, where the environment, company operations, and personal experiences all contribute to a sense of safety.

Introduction to the Safety Wheel

The main purpose of the tool is to understand customers' safety concerns to support customer-centric design by addressing diverse safety needs in various service contexts. The goal is to involve customers and stakeholders in the participatory design process to ensure their engagement and transparent communication. The developed tool is based on earlier research (Virkkula et al., 2024) and the user study presented in Figure 5.4.

Evaluation of the Tool

The purpose of the evaluation was to gain insight into the tool's practical user experience in the process of designing services. A focus group evaluation was conducted with 7 Finnish-speaking students ($M = 2$, $F = 5$) from the University of Lapland, representing different disciplines of media and design. Two groups of three to four participants used their earlier conceptual designs on regenerative tourism services to explore customer safety. Each group activated their concept using a predefined user profile. The objectives for the sessions were to assess the effectiveness

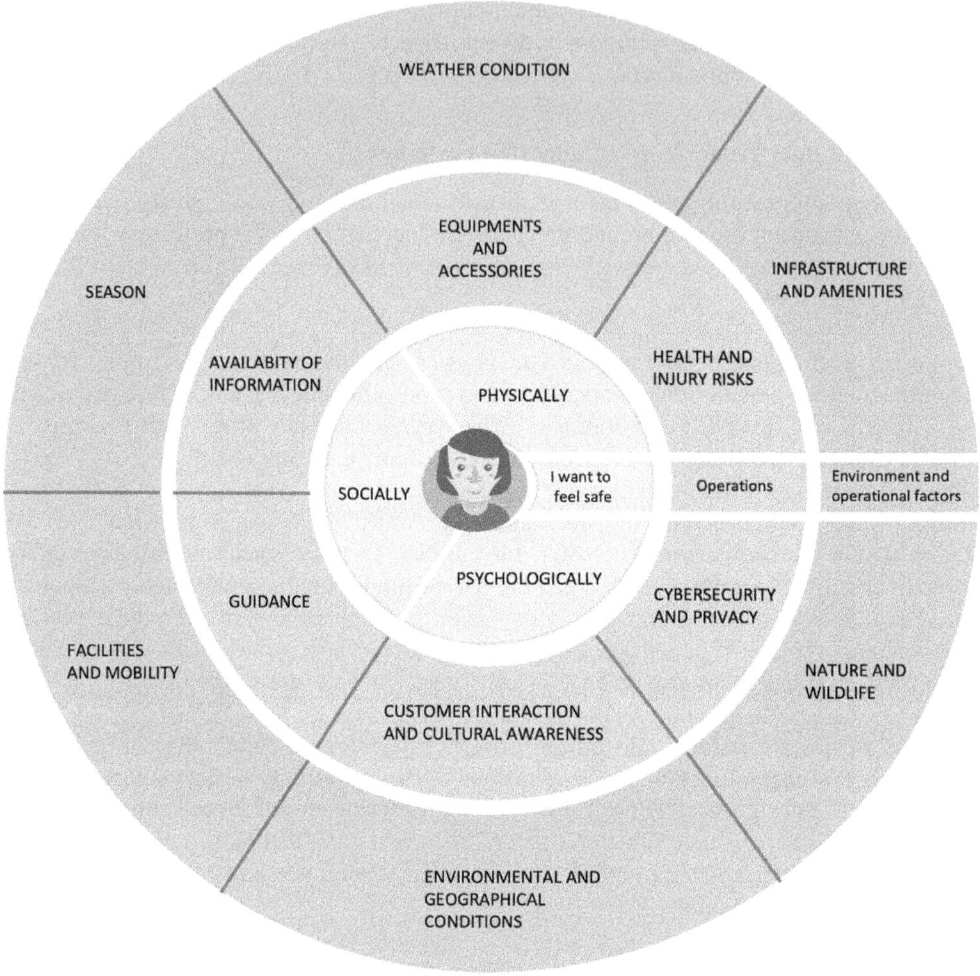

Figure 5.4 The structure of the tool is based on the focus group study.

of the safety tool, evaluate user experience, and examine engagement related to collaboration and structured discussion in a group.

In the evaluation, mixed methods of facilitating the group session were used. As shown in Figure 5.5, the first concept-based task for participants was to use the initial version of the tool to list important customer concerns related to their concept. The researcher observed users as they performed tasks and noted any difficulties or confusion. Users were encouraged to verbalize their thoughts while using the tool, providing insights into their interactions. In the second task, a blank template was provided to the groups to gather improvement ideas for the tool. Lastly, information was collected utilizing a questionnaire on individual participants' experiences, as well as their suggestions for improvement.

The collected data was analyzed using a qualitative mixed-methods approach, combining insights from the Safety Wheel tool, observation, templates, and questionnaires. This enabled a

Figure 5.5 Focus group session using the tool for reviewing the group's concept. Photograph by Minna Virkkula.

comprehensive understanding of the tool's effectiveness and user experience. Through thematic analysis, patterns were identified and recommendations for improvements were made.

Results from Evaluation

Observations and questionnaire feedback indicated that the overall functionality and usage of the tool received mostly positive responses, with only minor issues reported. In terms of supporting group interaction, the tool was seen to better suit its purpose, implying that it effectively facilitated collaboration among participants. Additionally, the tool was successful in capturing many safety concerns related to the concept, demonstrating its effectiveness in identifying potential customer safety issues in service concepts.

When asking participants how they experienced using the tool, the responses were mostly positive. They mentioned that the tool and the list of safety concerns helped them focus on various safety aspects that they had not previously considered. Examples of participants' comments regarding the use of the tool were: "*Eye opening, got to think of factors we did not think of before*". *For example,* "*how different age groups have different safety needs*" (P3) and "*Tool was comprehensive, and the list of safety concerns was helpful*" (P5). Although the content of the tool was perceived as comprehensive, the English language posed challenges for some participants, as mentioned in feedback: "*The English language was partly challenging*" (P5). Overall, the tool helped focus and direct the group's discussion broadly on service-related safety issues, as one of the participants commented: "*Interesting, it helped us to discuss in our group on different safety aspects of the service*" (P2).

Participants identified several challenges in understanding or applying the tool. The extensive text, particularly in the list of safety concerns, was overwhelming to some. Also, more detailed instructions on using the tool, such as titles and starting points, were needed. Feedback from the participants included "*I would have needed more detailed instructions for using the circle (titles, where to start, etc.)*" (P6), "*A lot of text, especially in the list of safety concerns. Maybe more graphical elements or visual guidance would be helpful*" (P3), and "*Fewer themes to make the content easier to comprehend, especially on the outer circle*" (P4).

As for additional resources needed to support safety design, participants suggest having a more detailed description of user profiles would be beneficial, as well as having a Finnish version of the tool. Participants also noted a tendency to focus on the "during the service" stage, mainly neglecting the "before the service" stage, where concerns differ. Guidance on addressing this issue would be supportive, as shown in these comments: "*More detailed description of user profile to start with*" (P6), "*We keep on thinking of the 'using the service' phase and did not concentrate before the usage of the service, where concerns are a bit different*" (P7).

Focus group participants recommend combining some of the higher-level safety themes when using fill-in templates, such as weather and season, to simplify the tool, as commented: "*Some themes could be combined together, like weather and season*" (P1). Especially the themes of the outer circle were not distinguishable enough. Reducing the number of themes would help users better understand the related aspects, as suggested by participants: "*Fewer themes to understand what aspects are related to them*" (P5).

There were no instructions for using the tool. Groups started discussing and utilizing the tool from the outer circle. Instructions to start from the customer's perspective would be useful. The focus group discussion was lively and centered around safety issues. Participants actively engaged in conversations, sharing their suggestions regarding various safety aspects of the service. A more spin-shaped tool could provide broader insights, as suggested in the group's fill-in template (i.e., G2): "*A more spin-shaped tool would provide more diverse results*" (see Figure 5.6). The theme "Facilities" was seen as a part of companies' operations and should be moved to the central circle, as suggested in the fill-in template (i.e., G1). Rephrasing the circles and themes, as well as combining the themes, was suggested in both group templates, and this activity provided valuable suggestions for improving the tool.

Collected recommendations for enhancing the tool's structure, content, and phrasing to make it more intuitive to use in a group are based on the evaluation:

1. **Simplifying and combining the themes:** Participants recommend focusing more on higher-level safety themes and combining items in the outer circle to simplify the tool. Reducing the number of themes helps users better understand the related aspects.
2. **Text-based content:** There were various suggestions for phrasing the circles. The suggested labels were "Personal Safety", "Business Operations", and "Operational Context".
3. **Guidance for using the tool and examples:** Provide clear instructions on how to use the tool. The guide should start from the center, emphasizing the user's perspective. Common safety concerns should be included as concrete usage examples.
4. **Naming the tool:** A few name proposals included "SafetySpinner", "Safety Radar", or "Roller".

Figure 5.6 Participants collaboratively brainstorm improvements using a fill-in template. Photograph by Minna Virkkula.

5. **Utilizing graphical elements and visualizations:** Participants suggested providing icons and other graphical cues, such as arrows pointing outward from the center circle, to guide users through the tool.
6. **Language support:** Offer multiple language options to ensure accessibility and eliminate language barriers.

Findings of the Implementation of the SafetySpinner Tool for Tourism Services

This case study presents the research process of developing the SafetySpinner, a novel tool for participatory safety design in tourism services. It focuses holistically on safety, filling a gap not addressed by existing service design methods. The SafetySpinner promotes collaboration, transparency, and customer-centricity, with a specific focus on customers' safety experience. Findings indicate that using the SafetySpinner was seen as appealing due to its ability to address a wide range of safety concerns comprehensively. The tool also enables empathy toward the customer, generating a new perspective on the identification of users' safety needs. The study demonstrates that the tool facilitates communication and open discussion within the group and brings together the views of customers and participants through visual means. Additionally, the tool emphasizes the importance of addressing safety concerns across user groups, promoting an inclusive approach. By integrating diverse perspectives and offering a visual aid for effective interaction, the tool fosters a strong focus on safety.

The contribution of the study is to introduce a new way of thinking about safety in services, with a practical tool that enhances customer knowledge, communication, and engagement in the design process. The tool addresses a broad scope of safety aspects and promotes a safety culture in services. The tool benefits from the collective insight of customers' safety concerns related to services and aids in concentrating on important aspects of safety by establishing a common understanding. This approach also enhances customer knowledge, mindset, and emotional potential, ensuring that expectations regarding safety are aligned with user needs.

Common challenges faced in co-creation practices of service innovations include ensuring effective communication, managing diverse stakeholder standpoints, and proposing solutions to navigate over these barriers, according to Holmlid et al. (2015). By integrating different perspectives through cooperative effort, a holistic view on safety is achieved, addressing safety concerns from the customer's point of view. Effective interaction and engagement foster a cooperative environment, allowing participants to contribute diverse perspectives and ensuring that various concerns are addressed.

Transparent communication is the foundation for trust and mutual understanding (Virkkula et al., 2024). Open communication can enhance customer experience and strengthen perceived service quality. Ultimately, this results in a new way of thinking about safety in services, more consistent service design, and a stronger commitment to user safety and well-being.

Discussion

The SafetySpinner tool emphasizes communication and a holistic focus on safety, ensuring a thorough walkthrough of customers' safety concerns among participants and aiding in building trust between the customer and the service provider. Different themes are visible and handled simultaneously throughout the service, from customers' information retrieval related to the service to collecting feedback.

The study conducted by Bąk and Szczecińska (2020) highlights shifts in demographics, including population aging, urbanization, and alterations in family compositions that differ across regions. These factors influence customer preferences and demand, which need to be considered in safety design. Addressing safety needs across user groups includes seniors' concerns, technology integration for younger users, and adapting solutions for diverse family structures. Acknowledging these factors in participatory design leads to more user-friendly, accessible, and successful outcomes. Additionally, there are notable trends, such as the heightened importance of personal safety and increased ethical considerations regarding sustainability. Supporting tools are essential for designing the safety experience in service design processes to achieve the desired outcome. Among these tools is a customer profile, which focuses on defining user needs across demographics, a factor that emerged in the study. Additionally, using a customer journey map supports identifying safety aspects throughout the entire service process.

Empathy and safety play important roles in participatory design, contributing to user-centric outcomes. Koskinen (2023) examines in his book how empathic and interpretive design can be integrated into participatory design methodology. Research demonstrates that applying user-centric design leads to functional, meaningful products and environments. Emphasizing social responsibility and empathy enhances understanding, meets user needs, and fosters collaboration. Participatory and inclusive design engages customers and stakeholders, while safety-focused approaches improve service quality. Additionally, aligning safety design with sustainability goals can expand the scope of the SafetySpinner tool for future developments. Integrating the safety aspect into participatory design ensures service user satisfaction.

Conclusion

In conclusion, this research explored the role of safety experience in services, aiming to develop a tool for participatory safety design for the tourism industry. By developing and evaluating the visual SafetySpinner tool, the study has demonstrated how collaborative design approaches can address holistic safety needs and share different perspectives through cooperative effort. The key insights indicate that the evaluated tool provided satisfactory functionality and significantly enhanced interaction among groups, supporting participant collaboration. By adopting the customer's perspective, the tool supports customer-centric service design and enhances overall service quality and reliability.

However, the limitations include not involving a wider participant group than design students, which may affect the applicability of the findings and limit the extent to which diverse perspectives are considered. Moreover, potential areas for future research should aim to identify stakeholders involved and their roles in implementing safety and evaluate the SafetySpinners' scalability and applicability to different kinds of services.

Ultimately, this research contributes to safety design for tourism services by providing a practical tool for participatory design, ensuring that customers' safety is aligned with their expectations. Overall, the structure of the tool aids in collaborative work, sharing perspectives, and fostering an emotional connection with customers. A revised version is under development based on study feedback, aiming not only to improve service safety but also to build customer trust and enhance the overall experience.

References

Almenara, A. P., Elich, J. H., & Saltiveri, T. G. (2024). Past, present and future of user-centred design methodologies that empathise with the user and generate better accessible experiences from theory to practice. In S. Sayago, M. Sendín, & M. Teixidó (Eds.), *Future perspectives on human-computer interaction research: Towards the year 2030* (pp. 17–43). Springer.

Bak, I., & Szczecinska, B. (2020). Global demographic trends and effects on tourism. University of Piraeus. International Strategic Management Association.

Barbe, D., & Pennington-Gray, L. (2020). Social media and crisis communication in tourism and hospitality. In Z. Xiang, M. Fuchs, U. Gretzel & W. Höpken (Eds.), *Handbook of e-Tourism* (pp. 1–27). Springer.

Bellato, L., Frantzeskaki, N., & Nygaard, C. A. (2023). Regenerative tourism: A conceptual framework leveraging theory and practice. *Tourism Geographies*, 25 (4), 1026–1046. https://doi.org/10.1080/14616688.2022.2044376

Bentley, T. A., Page, S. J., & Macky, K. A. (2007). Adventure tourism and adventure sports injury: The New Zealand experience. *Applied Ergonomics*, 38 (6), 791–796. https://doi.org/10.1016/j.apergo.2006.10.007

Blessing, L. T. M., & Chakrabarti, A. (2009). *DRM, a design research methodology*. Springer.

Chauhan, V. (2007). Safety and security perceptions of tourists visiting Kashmir, India. In J. S. Chen (Ed.), *Advances in hospitality and leisure* (pp. 3–17). Emerald Group Publishing Limited.

Cooper, D. (2022). Exploring the potential of e-tourism in enhancing accessibility for disabled and elderly travelers. *Journal of Policy Options*, 5 (3), 24–29.

Dickson, T. J. (2012). An introduction to risk, adventure and risk management. In T. J. Dickson & T. L. Gray (Eds.), *Risk manage-ment in the outdoors: A whole-of-organisation approach for education, sport and recreation* (pp. 1–24). Cambridge University Press.

Eizenberg, E., & Jabareen, Y. (2017). Social sustainability: A new conceptual framework. *Sustainability*, 9 (1), 68.

Ehn, P. (2017). Scandinavian design: On participation and skill. In D. Schuler & A. Namioka (Eds.), *Participatory design* (pp. 41–77). Taylor & Francis.

Gerling, K. M., Meiners, A.-L., Schumm, L., Rixen, J., Wolf, M., Yildiz, Z., Alexandrovsky, D., & Opp, M. S. (2024). An equitable experience? How HCI research conceptualizes accessibility of virtual reality in the context of disability. arXiv. https://doi.org/10.48550/arXiv.2411.04489

Gidebo, H. B. (2021). Factors determining international tourist flow to tourism destinations: A systematic review. *Journal of Hospitality Management and Tourism*, 12 (1), 9–17.

Gregory, J. (2003). Scandinavian approaches to participatory design. *International Journal of Engineering Education*, 19 (1), 62–74.

Holmlid, S., Mattelmäki, T., Visser, F. S., & Vaajakallio, K. (2015). Co-creative practices in service innovation. In R. Agarwal, W. Selen, G. Roos & R. Green (Eds.), *The handbook of service innovation* (pp. 545–574). Springer.

Huddart, D., & Stott, T. (2020). *What is adventure tourism?* Springer. https://doi.org/10.1007/978-3-030-18623-4_1

Karl, M., & Schmude, J. (2017). Understanding the role of risk (perception) in destination choice: A literature review and synthesis. *Tourism: An International Interdisciplinary Journal*, 65 (2), 138–155.

Koskinen, I. (2023). *Design, empathy, interpretation: Toward interpretive design research.* MIT Press.

Luong, V. H., Manthiou, A., Kang, J., & Nguyen, C. (2024). The building blocks of regenerative tourism and hospitality: A text-mining approach. *Current Issues in Tourism,* 27 (3), 361–380.

Mansfeld, Y., & Pizam, A. (Eds.). (2006). *Tourism, security and safety.* Routledge.

Matiza, T., & Slabbert, E. (2021). Tourism is too dangerous! Perceived risk and the subjective safety of tourism activity in the era of covid-19. *Geo Journal of Tourism and Geosites*, 36, 580–588. https://doi.org/10.30892/gtg.362spl04-686.

Muller, M. J., & Druin, A. (2003). Participatory design: The third space in HCI. In J. A. Jacko & A. Sears (Eds.), *The human-computer interaction handbook: Fundamentals, evolving technologies and emerging applications* (pp. 1051–1068). Lawrence Erlbaum Associates.

Morgan, D. (2021). Safety management in the adventure tourism industry. In J. Wilks, D. Pendergast, P. A. Leggat, & D. Morgan (Eds.), *Tourist health, safety and wellbeing in the new normal* (pp. 373–387). Singapore: Springer. https://doi.org/10.1007/978-981-16-5415-2_15

Narangajavana Kaosiri, Y., Callarisa Fiol, L. J., Moliner Tena, M. Á., Rodríguez Artola, R. M., & Sánchez García, J. (2019). User-generated content sources in social media: A new approach to explore tourist satisfaction. *Journal of Travel Research*, 58 (2), 253–265.

Norman, D. (2013). *The Design of Everyday Things.* (Revised and expanded edition). Basic Books.

Simonsen, J., & Robertson, T. (Eds.). (2012). *Routledge international handbook of participatory design* (1st ed.). Routledge. https://doi.org/10.4324/9780203108543

Sharma, A., & Sodani, P. (2024). Ethics in tourism: Responsibility toward balancing sustain- ability. In A. Sharma (Ed.), *International handbook of skill, education, learning, and research development in tourism and hospitality* (pp. 629–647). Singapore: Springer.

Spinuzzi, C. (2005). The methodology of participatory design. *Technical Communication*, 52 (2), 163–174.

Stephanidis, C., Salvendy, G., Antona, M., Duffy, V. G., Gao, Q., Karwowski, W., Konomi, S., Nah, F., Ntoa, S., Rau, P.-L. P., Siau, K., & Zhou, J. (2025). Seven HCI grand challenges revisited: Five-year progress. *International Journal of Human–Computer Interaction*, 41 (19), 11947–11995. https://doi.org/10.1080/10447318.2025.2450411

Taylor, S., Varley, P., & Johnson, T. (2013). *Adventure tourism: Meaning, experience and learning* (Vol. 36). Routledge.

Virkkula, M., Hokkanen, L., & Häkkilä, J. (2024). Making visitors feel safe-design students' perspectives on safety concerns when designing future tourism services for north. In *Global sustainable tourism conference academic symposium: Purposeful travel.* Purdue University Press.

Virkkula, M., Hokkanen, L., Häkkilä, J., & Aro, P. (2024). Improving the festival safety management through design approach. In *DMI: Academic design management conference* (pp. 815–824). Design Management Institute.

Xie, C., Zhang, J., & Morrison, A. M. (2021). Developing a scale to measure tourist perceived safety. *Journal of Travel Research*, 60 (6), 1232–1251.

Zou, Y., & Meng, F. (2020). Chinese tourists' sense of safety: Perceptions of expected and experienced destination safety. *Current Issues in Tourism*, 23 (15), 1886–1899.

Zou, Y., & Yu, Q. (2022). Sense of safety toward tourism destinations: A social constructivist perspective. *Journal of Destination Marketing & Management*, 24, 100708.

6 Participatory Citizen Engagement in the North

Exploring Green Transition Futures with Generative AI

Siiri Paananen, Rauni Äärelä-Vihriälä, Pigga Keskitalo, and Jonna Häkkilä

Introduction

Artificial intelligence (AI), particularly generative artificial intelligence (GenAI), is advancing rapidly (e.g., Ali et al., 2021), and adopting GenAI tools provides new means of citizen participation. This chapter explores how GenAI could support engaging citizens through co-design methods. The study focuses on engaging underrepresented groups, particularly youth and older residents, to strengthen their role in decision-making. We investigate how GenAI tools can facilitate futures literacy, the capability to critically evaluate and construct future scenarios (Ahvenharju & Pouru-Mikkola, 2022). Without inclusive methods, green transition policies can reinforce existing inequalities rather than advance social justice. Through two participatory workshops organized in Northern Finland, we address a key gap in the literature on the application of GenAI in participatory settings with diverse age groups in rural and northern contexts and examine how it can support futures consciousness.

Prior work has examined how immersive and visual methods, such as virtual reality and AI-generated content, can support reflections on green transition, particularly in natural environments (Nurmela et al., 2024). LC et al. (2024) involved the participants in using GenAI to visualize the results of climate change through a participatory process, highlighting reflection and emotional engagement. Young people's engagement with the future in the present moment, experienced as an affective atmosphere, enables thinking about and anticipating the future (Mandich et al., 2024). Chopra et al. (2022) demonstrated that participatory speculative design both explores community futures, such as local food resilience, and highlights their potential for engagement. The importance of context is emphasized by McDonagh and Tuulentie (2020), who note that decision-making and land use in the northern peripheral regions remain underrepresented in academic research. Prior research has sought to find ways to explore the perspectives of children and the elderly living in the North, which could be explored through the lens of sustainability, including discussing the possibilities of art-based activities such as painting and photography (Äärelä-Vihriälä et al., 2024). We extend this prior work by focusing on the exploration of creative tools for intergenerational engagement, using GenAI to create workshop materials. Thus, we ask: How can generative AI support participatory design and the exploration of green transition futures in northern communities while considering local cultural, ecological, and intergenerational perspectives?

Ultimately, our chapter contributes to the broader discussion of how emerging technologies like GenAI can support more inclusive, speculative, and democratic forms of engagement. We

DOI: 10.4324/9781003667872-8

respond to the call for rethinking engagement in innovative ways, especially through design and technology, by showing how young people in northern communities can participate in shaping future scenarios using GenAI. Our work advances academic understanding of integrating GenAI into participatory design and provides practical insights for community-based, intergenerational practice and education.

Related Work

Generative AI as a Design Tool

A systematic review of the role of GenAI in participatory design revealed that GenAI can have various roles in supporting the process, whether as a muse, co-pilot, or evaluator (van den Broek et al., 2024). The review uncovered that GenAI can enable facilitation, collaboration, and creation in participatory design, although critical stakeholder engagement can be lacking. GenAI can also function as a visionary design tool (Huang et al., 2023). Tang et al. (2024) conducted a study on two user groups, professionals in art and design fields and non-professionals, highlighting that GenAI tools can make creative processes more accessible for all. As GenAI tools rapidly evolve, the design field is in transition, with designers adopting, testing, and sometimes abandoning new tools (Kalving et al., 2024), while also navigating changes to their own agency. For example, Guo et al. (2023) found that designers who began with only vague ideas often expected GenAI to complement their concepts, but when the results were disappointing, they often needed to take back control of the creative process. At the same time, in studies with design students, GenAI has functioned as a supportive co-creator rather than a replacement for human creativity (Gong et al., 2024). A critical examination of GenAI, particularly in relation to Indigenous knowledge, is essential, as GenAI is not neutral; its training data often reproduce societal biases that marginalize Indigenous knowledge and frame it through Western perspectives (Worrell & Johns, 2024).

GenAI has shown potential for participatory design in visualizing landscapes and outdoor environments. For example, Guridi et al. (2025) used GenAI to design public parks and found that it demonstrated increased variation in ideas during the early stages of the process. Generating possible future landscapes for the green energy transition has involved the use of immersive VR and GenAI to compare natural environments with wind energy facilities environments (Nurmela et al., 2024). In their study, LC and Tang (2023) explored the role of GenAI tools in imagining climate futures in a workshop setting. They report that generative tools can be used to prototype future scenarios and envision designs that serve social purposes but also point out challenges in controlling GenAI outputs and selecting the precise keywords needed to achieve a particular vision.

Educational and Northern Perspectives on Participation

The North is not merely a geographical location but a complex socio-cultural and political space marked by unique conditions. These include sparse populations, long distances, fragile ecosystems, and often limited infrastructure, all of which impact community engagement and access to new technologies. In many northern and Arctic regions, such as Northern Finland, these challenges intersect with rich cultural diversity and Indigenous knowledge systems, including Sámi perspectives, which are often underrepresented in mainstream design and technology discourses (Keskitalo, 2010; Pietikäinen & Dufva, 2006). Importantly, northern regions often lie at national policy margins yet are central to global interests in climate, resources, and the green transition (McDonagh & Tuulentie, 2020), creating tension between local realities and external development plans. In Northern Finland, land use conflicts around wind energy projects and Indigenous rights clearly illustrate the stakes (Nysten-Haarala et al., 2021). Participatory design in these contexts must, therefore, critically address questions of justice, voice, and representation (Costanza-Chock, 2020).

Urban and digitally privileged areas often serve as testbeds for technologies like GenAI, yet significant gaps remain in understanding their use in peripheral communities. The value of these tools must be assessed beyond technical novelty, considering local meanings, cultural continuity, and situated knowledge (Larsen & Fondahl, 2015). While GenAI tools offer scalable and engaging methods for participatory design, their use must be rooted in local knowledge systems and ethical practices. Participatory design with GenAI can support pluralistic and place-based approaches to sustainability education and offer lessons for educators, designers, and policymakers working in similar contexts (Costanza-Chock, 2020). When integrated thoughtfully, GenAI technologies serve not only as creative tools but also as pedagogical means for exploring how knowledge is constructed and communicated, thereby supporting educational aims related to citizenship and sociotechnical awareness (Selwyn, 2019).

Empowerment can be seen as both a process and an outcome of participation (Zimmerman, 2000). In the Arctic, participatory approaches are particularly significant for communities facing marginalization and rapid change. Sámi education illustrates this well: empowerment often emerges when teachers engage in culture-based teaching design, collaborate in workshops, implement Indigenous activities, and reinterpret knowledge for their classrooms—processes that strengthen autonomy and confidence (Nutti, 2018). Similar participatory design practices in northern contexts can help reclaim agency, challenge dominant narratives, and create dialogic spaces (Björgvinsson et al., 2012), while bringing lived experience into conversation with new tools such as AI-generated images can open possibilities to imagine alternative, pluriversal futures (Escobar, 2018). Empowerment relates to voice, representation, and influencing imagined futures, framing the North as an active knowledge producer rather than a passive innovation recipient. Such approaches contribute to the decolonization of technological design by foregrounding local agency and epistemologies; including the local and cultural context is essential in decolonizing design practices (Paananen et al., 2022). In summary, we advocate a place-based, ethically grounded use of GenAI in participatory design that respects local specificities, embraces pluralism, and recognizes communities' power to shape their own sustainable futures.

Building on these contextual and ethical considerations, and as this research is situated in the educational context of Finland, we draw the framework from Ahvenharju and Pouru-Mikkola (2022), who define futures literacy as the skill to imagine and critically engage with futures. This skill is grounded in futures consciousness, which can be defined as the ability to understand, anticipate, prepare for, and embrace the future (Ahvenharju et al., 2021). Connecting the practice of futures literacy skills to experiences that students find meaningful increases their engagement in learning those skills (Bol & de Wolf, 2023). Futures literacy aims to prepare young people to be active, engaged participants in shaping the future. Using the Four Resources model, Häggström and Schmidt (2021) frame education as transformative and stress active, sensory, place-based literacies that deepen children's awareness, independent exploration, and narrative expression. We also consider the land-based nature of education in Northern Finland, particularly within Indigenous schooling, where learning is grounded in land, local culture, and community, deepening students' understanding of their environment (Gaup et al., 2024).

Methods

The Ethical Framework of the Study

This study adheres to ethical guidelines, particularly regarding research involving groups such as children, youth, and older participants, and conducting research in northern contexts that include Sámi communities and perspectives. Ethical approval was obtained from the university's review board, aligned with Finland's national guidelines on protecting research participants. In participatory research conducted in Northern Finland, particular ethical attention is crucial due to diverse local communities, including Sámi, the Indigenous people of the region. While this study did not specifically target the Sámi community, ethical considerations were informed by best practices for research in Northern and Indigenous contexts (Heikkilä et al., 2024), emphasizing respect for local knowledge and cultural practices. Special care was taken with children and youth to ensure safety, dignity, and age-appropriate self-expression (Phelan & Kinsella, 2013). The workshops aimed to create an inclusive and respectful environment where participants of different ages could share their views openly and participate in their own way. The study followed ethical principles of informed consent, voluntary participation, and data privacy. All participants, and in the case of children, their guardians, provided informed consent. Participants were informed of their right to withdraw from the study at any time without consequences. No special category data and only minimal personal data were collected. All data were pseudonymized and handled following GDPR (Regulation (EU) 2016/679). Data were stored securely and managed according to the university's data management guidelines, ensuring transparency and ethical integrity in handling qualitative research materials (Tracy, 2010). In addition, we followed established principles for ensuring the reliability and rigor of research processes throughout the study (Bryman, 2018).

As we conduct our research in the context of climate change and the green transition, it is important to note that AI requires significant natural resources, particularly water and electricity, to function. Using AI tools in green transition research has environmental impacts, including energy consumption and ecological footprint (Bhaskar & Seth, 2024). Therefore, developing sustainable GenAI practices is crucial for a sustainable future.

Workshop Design and Facilitation

The workshops aimed to explore the experiences of children, youths, and elderly people regarding the justice of the green transition in Northern Finland and to investigate the potential and effectiveness of art-based approaches in examining intergenerational perspectives within the framework of sustainability research. The workshops used an art-based participatory method in which participants expressed their own perspectives through AI-generated illustrations. Art-based participatory research methods have been widely used in youth studies, as they help engage young people in the research process. However, participatory and creative methods also face practical constraints, such as institutional barriers and time management, which can limit the depth of participation by young people (Water et al., 2020). We addressed this by designing workshop timelines to ensure that voices from each session contributed to a coherent overall picture. To capture diverse responses and to test GenAI's adaptability, we used two workshop formats: a small intergenerational group and a larger youth-focused event.

The workshops were held in Northern Finland (2024–2025) and facilitated in Finnish. The first one was intergenerational, involving four elders and eight children/youths at a local school. No specific demographics were collected due to ethics considerations. The focus was

Figure 6.1 The second author leads the discussion on the GenAI image (OpenAI DALL·E, via ChatGPT, model GPT-4o). Photograph by Henri Mella.

on dialogue on local issues with the participants. Data from the first workshop comprised four audio recordings of group discussions around a printed GenAI image (Figure 6.1), handwritten notes on post-its, and researchers' diary entries.

The second workshop was part of a youth-focused workshop day attended by 130 middle school students who visited the university service design laboratory in smaller groups of approximately 15 students, spending about 30 minutes per group. Data included digital annotations made by youths on the smart screen version of the GenAI image, whiteboard annotations, and researchers' diaries. Unlike the manual annotations on paper in the first workshop, the second workshop used participants' digital annotations directly on the smart screen (Figure 6.2). At the end, participants gave general comments on the GenAI visuals.

As the input for generating the first image, we used information from Finland's Scenarios 2030: Lapland Edition (Tulevaisuuden Lappi: Suomi Skenaariot 2030), which discusses possible future directions for Finland and Northern Finland. The image was created using ChatGPT based on a scenario text retrieved from the Lapland Regional Council's website in April 2024 (https://www.lapinliitto.fi/) and was used in both workshops as a visual representation of a possible future (Figure 6.3). The second GenAI image used in the second workshop included an AI-generated story, based on researcher prompts informed by local conditions, about a local boy, Eetu, struggling to fish as the ice was weaker than usual due to hydropower effects, and an image based on this AI-generated story generated in Adobe Firefly (Figure 6.3).

Analytical Framework and Coding

The data were analyzed by the first two authors, using the dimensions of *futures consciousness:* Time Perspective, Agency Beliefs, Openness to Alternatives, Systems Perception, and Concern for Others (Ahvenharju et al., 2021). During coding, two additional categories emerged: Image Evaluation (straightforward observations about image content) and Gen Z Slang (youth-specific online language and humor). This coding framework enabled us to examine how participants critically engaged with sustainability through the speculative visuals. The final codebook is shown in Table 6.1.

Figure 6.2 Participants collaborated after reading a story and seeing a GenAI image. Photograph by Henri Mella.

Figure 6.3 The two GenAI images are the future vision of the city (left) and a boy fishing (right). (Left: OpenAI DALL·E (via ChatGPT, model GPT-4o). Right: Adobe Firefly, Firefly Image 3. Prompts by authors.)

Table 6.1 Codebook used in the analysis

Category	*Code*
Time perspective	A1
Agency beliefs	T1
Openness to alternatives	AV1
Systems perception	S1
Concern for others	V1
Image evaluation	K1
Gen Z Slang	Z1

Results

Next, we present the research findings according to the dimensions of futures consciousness. These dimensions provide a comprehensive framework for understanding how participants relate to and interpret possible futures, highlighting both individual and collective aspects of futures literacy.

Realism, Agency and Concern

The time perspective theme focused largely on how realistic participants found the AI-generated scenario. At the core of time perspective is the human ability to be aware of the past, present, and future and to understand how events are connected and follow one another over time (Ahvenharju et al., 2021). Since 2030 is no longer in the distant future, one participant called the visualized scenario "An overly idealized future" and, reflecting on its feasibility, commented that it is "Unlikely that Lapland would already look like this in 2030." Some children found the vision unrealistic, noting that "Nothing can change that much in 5 years," and some others described it as "too modern." Safety concerns arose, as illustrated by comments such as "Does not look like our city or feel safe," and "AI will kill us in Lapland one fine day:)." Also, the participants mentioned some of the buildings and locations, such as a library, the city center, and the local landmark bridge, and were reflecting on how the current locations and landmarks could be located in the future visions, as illustrated by the question "Where is the ice rink?". About the change in landscape, a child was wondering how a river could change like that. The discussions between the elderly and children involved local places and traditions, such as an elder wondering: "Yeah, so where exactly is the [local hill's name]?" and a child pointing out, "A kind of traditional feel—over there it is very modern—would be nice to preserve a certain traditional character."

Many agency-related comments referred to the AI-generated character, Eetu, the boy who is no longer able to fish due to changing river conditions shown in the image and story. Participants reflected on Eetu's options, discussing both the environmental constraints and the imaginable individual actions. People are active agents who shape how the future unfolds, as there is no predetermined future that humans could simply discover (Ahvenharju et al., 2018). Most responses involved finding a new fishing spot or addressing the issue. For example, children often proposed that Eetu should "Find a new fishing spot"; a few also suggested making a complaint to the city or taking more drastic measures to fight the issue, as illustrated by the comment "Blow up the hydroelectric power plant". These suggested courses of action for Eetu demonstrated proactive agency, which is one form of future-oriented engagement (Ahvenharju et al., 2018). Other comments related to the agency category and the city's vision of 2030 included the children taking personal initiative, as in "I will move away if it looks like this." An elder commented on the future vision: "…I would go cast my fishing rod there. I just started the season yesterday. I got a boat there, too, but I am waiting for the flood to pass," relating the workshop topics to their own experiential knowledge.

Relatively few comments coded in the category Concern for Others were identified, mainly ones related to the image of the boy fishing alone and to the 2030 city. The boy was perceived as lonely, sad, and even depressed, and the image elicited feelings of sadness and longing for them, as illustrated by the comment "Elicits a sense of wistfulness," but was also found "Relatable" for some. In the intergenerational workshop, the future city sparked discussions on green energy and nature values. One child said, "For me, it is important that there are water and trees – it brings a good feeling. However, I would not necessarily need those wind turbines.

They probably do not do any harm, but I would not necessarily need them." An elder similarly noted, "The built environment looks quite overwhelming—there is much green, but untouched nature only seems to exist further away."

Mixed Reactions and Energy Production Discussion

Openness to alternatives involved both openness to experience and critical thinking. Here, many comments on all the photos were found, with mixed reactions to the photos: some found them unrealistic, others expressed liking them. Skepticism was mainly directed at the cityscape, with comments like "not possible," "fake," "unrealistic," and "ugly." In contrast, the picture of Eetu was seen as realistic and natural, described as "real," a "peaceful place," "atmospheric," and even "boring." Tall buildings and dense environments drew negative reactions, described as "cramped," "buildings too tall," and "not enough space." Some found the cityscape "overwhelming" or "anxious," especially compared to more open, natural surroundings. One child noted the abundant greenery, saying, "…got trees even in weird places, on top of buildings and next to them—I do not really see anything bad about that." Many general comments across photos fit Openness to alternatives. Some expressed appreciation, calling the visions "cool," "exciting," or "the best," while others showed disbelief or irony, describing the images as "fake," "unrealistic," or "wishful thinking," with some explicitly questioning their feasibility or authenticity. Despite skepticism, one participant said the images "Look real from a distance." In several cases, reactions were brief emotional responses such as "wow," "scary," or "do not like it."

Systems perception appeared mainly in comments on green energy, nature, city planning, and innovations, linking these to broader systemic issues found both in the future city and the image of the boy. A young person wrote, "Wind power is a good thing!" and another noted that "Local energy production" exists. With many renewables shown, participants said, "A positive thing is emission-free energy production," and on the current status, "Do we even have local energy production? Not all of it is even used. It is a huge destruction when there are lots of hydropower plants that destroy rivers." A child pointed out, "Maybe it is about how future inventions and things will come and at what pace," showing reflection on larger frameworks of development.

Evaluating Images and Using Youth Slang

The participants also provided straightforward observations, simply pointing out what they saw without necessarily evaluating it, listing elements like "boat," "electric car," "many trees," or "solar panels on the ground." Some questioned unclear details, such as "What is that?", "blurry," or "looks like AI." The various spherical buildings caught particular attention, with comments like "Supermarket (ball)" or simply "What ball?" indicating both recognition and confusion about these unfamiliar building shapes. Participants also paid attention to features that revealed the image as AI-generated. One noted a distorted detail in a fishing reel (Figure 6.4) with an arrow and a comment: "noticed it is AI." Overall, these comments show descriptive engagement, identifying elements, naming them, and occasionally questioning things that seemed odd or unfamiliar.

In the second workshop, the youth used their own language, which is often used online, such as in memes, leading us to form a separate category. Each task contained online and youth language references, such as blaming fishing problems on a "skill issue" (jokingly referring to user error) and calling the boy a "sigma lone wolf" (self-reliant type). Other youth slang included "locked in ninja" (very focused) and "cap" (not true). While these did not always provide much

Figure 6.4 Participants' drawings over the GenAI image.

direct input for the image contents, they provided researchers with information on how the youth used their own vocabulary to describe the contents of the workshop. The participants did not explain these terms; the authors added the glosses for readability.

Researchers' Observations

Facilitators observed some groups critically questioning the desirability and realism of future scenarios, while others used the images as a springboard for imaginative thinking, suggesting improvements, alternatives, or personal wishes for the future of the city. This diversity indicated that the method supported both critical thinking and creativity. Although the AI's vision of the city was not entirely accurate, recognizable elements such as a river and evergreen trees helped to anchor the imagined futures in a familiar landscape. This partial familiarity supported participants in making connections to their own environment and experiences. The participants also identified possible familiar locations.

In workshop two, the school groups of two to three students discussed the GenAI materials collaboratively. These small group dynamics appeared to facilitate more focused conversations, allowing participants to reflect on how the depicted futures might affect their daily lives or communities. The images prompted comments not only on environmental issues, such as climate change or pollution, but also on local infrastructure, urban planning, and social life. The interactive tasks also encouraged the participants to comment on each other's thoughts, whether by writing down "agree" or discussing it verbally. In sum, the results showed that imperfect AI images with localized cultural elements can stimulate meaningful dialogue. The combination of visual prompts, group discussion, and a shared focus on local futures proved effective in fostering engagement. While the cityscape was often seen as unrealistic, the picture of the boy seemed to evoke many feelings. This shows that AI-generated material can connect with participants' emotions and bring out their perspectives. Overall, the AI images served as effective discussion starters for youth and elders. The tasks were simple, enabling broad participation through discussion, writing, or drawing.

Discussion

Visuals Supporting Intergenerational Dialogue

Notably, the visual aspect of the workshop helped bridge generational gaps. Elderly participants occasionally shared memories related to specific locations or past environmental changes, which in turn prompted discussions among the youth. These exchanges enriched the workshop atmosphere and highlighted the value of diverse perspectives in future planning and sustainability. The ownership of the plans was noticed when the youth used their own language to discuss the images, as young people construct their identity and sense of community using English. As Leppänen (2007) shows, Finnish-speaking youth use English creatively to construct identities and signal community affiliation.

A key takeaway was the realization that GenAI can function as a powerful tool to ground abstract policy-level ideas in concrete, localized imagery, a format that people could emotionally and cognitively respond to. Rather than relying solely on textual descriptions from municipal green transition plans, the GenAI translations into visual futures made it easier for participants to imagine their own neighborhoods and living environments in a transformed state. This localization aspect, bringing the discussion closer to one's own environment, was seen as crucial in terms of both engagement and empowerment. Workshop participants often expressed surprise at seeing familiar buildings and landscapes reframed as part of future scenarios. The photos may bring the participant a feeling that these were not just someone else's plans anymore, but theirs to think about. This shift in ownership over the future narrative is a promising direction for citizen participation in sustainability planning (Kerr et al., 2022). Literacy practices and pedagogical choices should support the development of critical thinking about the relationship between humans and nature (Häggström & Schmidt, 2021).

AI as a Flexible Tool

Another strength of using GenAI was the flexibility and adaptability it provided. The workshop content could be easily "customized" to suit the needs and identities of different groups and places. GenAI tools allowed for the generation of scenarios that felt locally relevant—something that static, top-down communication methods often fail to achieve. The ability to quickly generate visionary scenarios and landscapes and flexibly modify them is a strength that is well-suited when preparing material for participatory workshops. However, we also confirm the challenge of controlling the appearance or disappearance of visualized items in generated views reported in prior art (LC & Tang, 2023). It was also apparent that GenAI cannot replace everything; facilitators were needed to organize, guide, and engage the participants, showing the need for social and emotional work pointed out by Selwyn (2019). The social aspects of innovation and development should not be forgotten in the sociotechnical reality we live in (Costanza-Chock, 2020).

It is also worth noting that this flexibility connects with the idea of culturally responsive design, here applied to northern contexts. However, we also acknowledge that GenAI is not a neutral tool, as it can reflect the training data and design choices. Participants also noticed details that seemed unrealistic or confusing, reminding us that these tools both shape and limit how futures are shown. Future AI applications must consider whose perspectives are included and whose are missing. The prior art has documented bias in GenAI visualizing everyday life environments, where, e.g., Western influence and attitudes are shown in created visuals of India (Jones et al., 2024). If participation is to be genuinely inclusive, the tools used should be adaptable to different knowledge systems and values, not just pushing one predefined version of the future. In this way, GenAI can even be seen as a potentially decolonizing tool when used with care, allowing space for multiple ways of knowing and imagining. Such an approach challenges the dominance of mainstream epistemologies and affirms the value of Indigenous worldviews. It invites collaboration rather than replacement and positions GenAI not as a neutral technology but as one that can either reinforce or resist existing power structures, depending on how and by whom it is shaped. This aspect of GenAI should be taken into consideration when planning citizen participation through GenAI tools. One important approach is to train teachers to use GenAI in alignment with Indigenous cultural perspectives and identities, enabling a deeper understanding of how the outputs can reflect local educational and cultural needs (Fausto et al., 2024).

Implications for Practice

Building on our findings, we outline implications for the consideration of other researchers and educators working on GenAI and participatory workshops. First, customizing the prompts to reflect local landscapes, discussions, and culture can make the engagement meaningful for the participants. Second, GenAI can be used to invite critique, humor and emerging storytelling, rather than presenting polished future visions. Third, planning a diverse set of ways of participating, whether it is by writing, drawing, or discussion, can help with engagement. Finally, being transparent about GenAI limitations and exploring ways to minimize environmental impact are important for more responsible, sustainable use of these tools.

Conclusions

This study explored the use of GenAI in participatory workshops to engage diverse community members, particularly children, youth, and elders, in imagining local futures shaped by climate change and sustainability transitions. Our study highlights how localization and visual storytelling can shift participants' perceptions of agency, ownership, and relevance in relation to complex future scenarios. However, our evidence comes from two workshops in a single setting, so the findings should be treated as exploratory rather than generalizable.

As a key finding, we highlight the context: when participants saw familiar buildings and landscapes in the GenAI images, the scenarios felt more immediate and personally relevant. This localization function supported emotional engagement and opened conversations that might otherwise have remained abstract or disengaged. Participants expressed surprise and curiosity, suggesting that the tools used can provoke reflection and dialogue. The workshops revealed how creative technologies serve as mediating tools between generations. The GenAI images triggered memories from elderly participants, while younger participants engaged with humor, slang, or critical reflection. Rather than imposing a singular vision, the approach encouraged co-creation, as participants brought their values, memories, and aspirations into the conversation. This points toward culturally responsive design, especially relevant in Northern and Indigenous contexts. Our findings also highlight the value of speculative imagination in sustainability education and planning. Futures literacy enables participants to actively engage with possible futures. By exploring future scenarios through a visual-narrative lens, participants engaged in futures thinking not as passive recipients but as active co-designers. This shift in narrative ownership, from institutional planners to local citizens, is particularly promising for developing long-term commitment to sustainability goals.

However, questions remain about the ethical use and representation in AI-mediated processes. In conclusion, GenAI can be a powerful enabler of inclusive sustainability transitions when implemented with cultural sensitivity, ethical awareness, and participatory intent. The value lies not in the technology itself but in how it is used, opening space for dialogue, imagination, and collective responsibility. Further research should explore how GenAI can support culturally grounded, community-based future-making, helping empower communities to shape their own sustainable futures.

Acknowledgments

This research has been supported by the project "Reconceptualizing Boundaries Together Towards Resilient and Just Arctic Future(s) (REBOUND)" (2023–2026), funded by the Strategic Research Council (SRC) established within the Research Council of Finland, grant numbers 358482, 358499, and 358498.

References

Äärelä-Vihriälä, R., Paananen, S., Häikiö, T. K., Keskitalo, P., Lohilahti, M. L., & Häkkilä, J. (2024). Engaging northern children's perspectives on sustainability: collaborative art-based methods in education. In T. Jokela, M. Huhmarniemi, & K. Burnett (Eds.), *Relate North: New Genre Arctic Art Education: Beyond Borders* (pp. 180–206). InSEA Publications.

Ahvenharju, S., Lalot, F., Minkkinen, M., & Quiamzade, A. (2021). Individual futures consciousness: Psychology behind the five-dimensional futures consciousness scale. *Futures, 128,* 102708. https://doi.org/10.1016/j.futures.2021.102708

Ahvenharju, S., Minkkinen, M., & Lalot, F. (2018). The five dimensions of futures consciousness. *Futures: The Journal of Policy, Planning and Futures Studies*, *104*, 1–13. https://doi.org/10.1016/j.futures.2018.06.010

Ahvenharju, S., & Pouru-Mikkola, L. (2022). Tulevaisuustietoisuus ja tulevaisuuslukutaito – kuinka kohtaamme, kuvittelemme ja käytämme tulevaisuutta [Futures consciousness and futures literacy – How we encounter, imagine, and use the future]. In H. K. Aalto, K. Heikkilä, P. Keski-Pukkila, M. Mäki, & M. Pöllänen (Eds.), *Tulevaisuudentutkimus tutuksi – Perusteita ja menetelmiä* (pp. 389–399). Tulevaisuuden tutkimuksen seura. https://urn.fi/URN:NBN:fi-fe2022081154412

Ali, S., DiPaola, D., Lee, I., Hong, J., & Breazeal, C. (2021). Exploring generative models with middle school students. In *Proceedings of the 2021 CHI Conference on Human Factors in Computing Systems* (pp. 1–13). Association for Computing Machinery. https://doi.org/10.1145/3411764.3445226

Bhaskar, P., & Seth, N. (2024). Environment and sustainability development: A ChatGPT perspective. In J. Singh, S. Goyal, R. Kumar Kaushal, N. Kumar, & S. Singh Sehra (Eds.), *Applied Data Science and Smart Systems* (pp. 54–62). CRC Press. https://doi.org/10.1201/9781003471059

Björgvinsson, E., Ehn, P., & Hillgren, P. A. (2012). Agonistic participatory design: Working with marginalised social movements. *CoDesign*, *8*(2–3), 127–144. https://doi.org/10.1080/15710882.2012.672577

Bol, E., & de Wolf, M. (2023). Developing futures literacy in the classroom. *Futures: The Journal of Policy, Planning and Futures Studies*, *146*, 103082. https://doi.org/10.1016/j.futures.2022.103082

Bryman, A. (2018). *Samhällsvetenskapliga metoder [Methods in Social Sciences]* (3rd ed). Liber.

Chopra, S., Clarke, R. E., Clear, A. K., Heitlinger, S., Dilaver, O., & Vasiliou, C. (2022). Negotiating sustainable futures in communities through participatory speculative design and experiments in living. In *Proceedings of the 2022 CHI Conference on Human Factors in Computing Systems* (pp. 1–17). Association for Computing Machinery. https://doi.org/10.1145/3491102.3501929

Costanza-Chock, S. (2020). *Design Justice: Community-Led Practices to Build the Worlds We Need.* MIT Press. https://doi.org/10.7551/mitpress/12255.001.0001

Escobar, A. (2018). *Designs for the Pluriverse: Radical Interdependence, Autonomy, and the Making of Worlds.* Duke University Press.

Fausto, I. R. d. S., Leta, F. R., & Braz, R. M. M. (2024). Application of artificial intelligence in indigenous pedagogy: A case study in Basic Education. *Revista Ibero-Americana de Estudos em Educação, 19*(esp. 3), e19321. https://doi.org/10.21723/riaee.v19i00.1932102

Gaup, E. U., Keskitalo, P., Korte, S.-M., & Moore, S. (2024). Sámi teacher education emphasising land-based approaches for Indigenous rights. In A. Osler, & B. Goldschmidt-Gjerløw (Eds.), *Nordic Perspectives on Human Rights Education: Research and Practice for Social Justice* (pp. 169–181). Routledge. https://doi.org/10.4324/9781003340676-14

Gong, Z., Paananen, S., Nurmela, P., Gonçalves, M., Georgiev, G. V., & Häkkilä, J. (2024). AI role in ideation for design creativity enhancement. In Y. S. Kim, Y. Nomaguchi, C.-H. Chen, X. Xin, L. Hu, & M. Wang (Eds.), *DS 136: Proceedings of the Asia Design and Innovation Conference 2024* (pp. 46–54). The Design Society.

Guo, X., Xiao, Y., Wang, J., & Ji, T. (2023). Rethinking designer agency: A case study of co-creation between designers and AI. In D. De Sainz Molestina, L. Galluzzo, F. Rizzo, & D. Spallazzo (Eds.), *IASDR 2023: Life-Changing Design,* 9–13 October, Milan, Italy. Design Research Society. https://doi.org/10.21606/iasdr.2023.478

Guridi, J. A., Cheyre, C., Goula, M., Santo, D., Humphreys, L., & Shankar, A. (2025). Image generative AI to design public spaces: A reflection on how AI could improve co-design of public parks. In *Proceedings of the 2023 CHI Conference on Human Factors in Computing Systems*, Article 7 (14 pages). Association for Computing Machinery. https://doi.org/10.1145/3656588

Häggström, M., & Schmidt, C. (2021). Futures literacy – To belong, participate and act!: An Educational perspective. *Futures: The Journal of Policy, Planning and Futures Studies*, *132*, 1. https://doi.org/10.1016/j.futures.2021.102813

Heikkilä, L., Kuokkanen, R., Lehtola, V. P., Magga, P., Magga, S. M., Näkkäläjärvi, J., Valkonen, S., & Virtanen, P. K. (2024). *Saamelaisia koskevan tutkimuksen eettiset ohjeet Suomessa = Sámiide guoski dutkama ehtalaš rávvagat Suomas = Sämmilijd kyeskee tutkâmeettisiih ravvuuh Suomâst = Sä'mmlaid kuõskki tu'tǩǩummuž eettla vuä'ppõõzz Lää'ddjânnmest = Ethical guidelines for research involving the Sámi people in Finland.* University of Oulu. https://urn.fi/URN:NBN:fi:oulu-202405294076

Huang, Y. C. J., Wensveen, S. A., & Funk, M. (2023). Experiential speculation in vision-based AI design education: Designing conventional and progressive AI futures. *International Journal of Design*, *17*(2), 1–17. https://doi.org/10.57698/v17i2.01

Jones, M., Raju, D. K., Pearson, J., Reitmaier, T., Robinson, S., & Majhi, A. (2024). Beyond 'slumming it'—AI and the real lives of global South communities. *Interactions*, *31*(6), 20–27. https://doi.org/10.1145/3696019

Kalving, M., Colley, A., & Häkkilä, J. (2024). Where AI and design meet-designers' perceptions of AI ools. In *Proceedings of the 13th Nordic Conference on Human-Computer Interaction* (pp. 1–8). ACM. https://doi.org/10.1145/3679318.3685388

Kerr, S., Krogh, U., & Riede, F. (2022). Experimental participatory methodology brings local pasts to contemporary climate action. *Climate Action*, *1*(Article 5). https://doi.org/10.1007/s44168-022-00006-5

Keskitalo, P. (2010). *Saamelaiskoulun kulttuurisensitiivisyyttä etsimässä kasvatusantropologian keinoin [Searching for Cultural Sensitivity of Sámi School Through Educational Anthropology].* Doctoral dissertation. Guovdageaidnu, Norway: Sámi allaskuvla.

Larsen, J. N., & Fondahl, G. (Eds.). (2015). *Arctic Human Development Report: Regional Processes and Global Linkages.* Nordic Council of Ministers. http://dx.doi.org/10.6027/TN2014-567

LC, R., Liu, S., Hendra, L. B., & Fu, K. (2024). TIME ENOUGH: Generative AI visions of climate change as cave paintings of the future. In *Proceedings of the 16th Conference on Creativity & Cognition* (pp. 608–613). Association for Computing Machinery. https://doi.org/10.1145/3635636.3672190

LC, R., & Tang, Y. (2023). Speculative design with generative AI: Applying stable diffusion and ChatGPT to imagining climate change futures. In *Proceedings of the 11th International Conference on Digital and Interactive Arts* (Article 36, 8 pages). Association for Computing Machinery. https://doi.org/10.1145/3632776.3632827

Leppänen, S. (2007). Youth language in media contexts: Insights into the functions of English in Finland. *World Englishes*, *26*(2), 149–169. https://doi.org/10.1111/j.1467-971X.2007.00499.x

Mandich, G., Satta, C., & Cuzzocrea, V. (2024). Feeling the future: An exploration into studying youth futures. *Futures: The Journal of Policy, Planning and Futures Studies*, *155*, 103299. https://doi.org/10.1016/j.futures.2023.103299

McDonagh, J., & Tuulentie, S. (Eds.). (2020). *Sharing Knowledge for Land Use Management: Decision-Making and Expertise in Europe's Northern Periphery.* Edward Elgar Publishing. https://doi.org/10.4337/9781789901894

Nurmela, P., Etto, J., & Häkkilä, J. (2024). Exploring VR as a method to study user perceptions of challenges in green transition technology in nature. In *Proceedings of the International Conference on Design, Learning, and Innovation* (pp. 51–68). Springer Nature Switzerland. https://doi.org/10.1007/978-3-031-85663-1_4

Nutti, Y. J. (2018). Decolonizing indigenous teaching: Renewing actions through a critical Utopian action research framework. *Action Research*, *16*(1), 82–104. https://doi.org/10.1177/1476750316668240

Nysten-Haarala, S., Joona, T., & Hovila, I. (2021). Wind energy projects and reindeer herders' rights in Finnish Lapland: A legal framework. *Elementa: Science of the Anthropocene*, *9*(1), Article 00037. http://doi.org/10.1525/elementa.2020.00037

Paananen, S., Suoheimo, M., & Häkkilä, J. (2022). Decolonizing design with technology in cultural heritage contexts – Systematic literature review. In *Congress of the International Association of Societies of Design Research* (pp. 1839–1855). Springer. https://doi.org/10.1007/978-981-19-4472-7_119

Phelan, S. K., & Kinsella, E. A. (2013). Picture this… safety, dignity, and voice—Ethical research with children: Practical considerations for the reflexive researcher. *Qualitative Inquiry*, *19*(2), 81–90. https://doi.org/10.1177/1077800412462987

Pietikäinen, S., & Dufva, H. (2006). Voices in discourses: Dialogism, critical discourse analysis and ethnic identity. *Journal of Sociolinguistics*, *10*(2), 205–224. https://doi.org/10.1111/j.1360-6441.2006.00325.x

Selwyn, N. (2019). *Should Robots Replace Teachers? AI and the Future of Education.* Polity Press.

Tang, Y., Zhang, N., Ciancia, M., & Wang, Z. (2024). Exploring the impact of AI-generated image tools on professional and non-professional users in the art and design fields. In *Companion Publication of the 2024 Conference on Computer-Supported Cooperative Work and Social Computing* (pp. 451–458). Association for Computing Machinery. https://doi.org/10.1145/3678884.3681890

Tracy, S. J. (2010). Qualitative quality: Eight "big-tent" criteria for excellent qualitative research. *Qualitative Inquiry*, *16*(10), 837–851. https://doi.org/10.1177/1077800410383121

van den Broek, S., Sankaran, S., de Wit, J., & de Rooij, A. (2024). Exploring the supportive role of artificial intelligence in participatory design: A systematic review. In *Proceedings of the Participatory Design Conference 2024: Exploratory Papers and Workshops-Volume 2* (pp. 37–44). Association for Computing Machinery. https://doi.org/10.1145/3661455.3669868

Water, T., Payam, S., Tokolahi, E., Reay, S., & Wrapson, J. (2020). Ethical and practical challenges of conducting art-based research with children/young people in the public space of a children's outpatient department. *Journal of Child Health Care*, *24*(1), 33–45. https://doi.org/10.1177/1367493518807318

Worrell, T., & Johns, D. (2024). Indigenous considerations of the potential harms of generative AI. *Agora (Melbourne*, Vic.), *59*(2), 33–36.

Zimmerman, M. A. (2000). Empowerment theory: Psychological, organizational and community levels of analysis. In J. Rappaport & E. Seidman (Eds.), *Handbook of Community Psychology* (pp. 43–63). Springer. https://doi.org/10.1007/978-1-4615-4193-6_2

7 Rethinking Civic Engagement

Chorong Kim and Ki-Young Nam

Introduction

In the evolving landscape of civic engagement, the intersection of design thinking and policy development presents an innovative framework for understanding and addressing the diverse needs of citizens. Design thinking, based on user needs, can be translated into 'citizen-centred' approaches in the context of policymaking, offering distinctive pathways for enhancing civic participation. The design thinking approaches have power beyond traditional design—as in design for social innovation—aligning seamlessly with the objectives of civic engagement, as they can facilitate policymakers to delve deeper into the genuine needs of citizens rather than merely viewing participation as a procedural obligation for democratic legitimacy. By integrating technology-mediated tools, these approaches can become even more powerful in fostering deeper connections between citizens and policymakers.

Historically, civic participation has often been treated as a formality essential for legitimizing policies rather than as a fundamental mechanism for understanding citizen needs. This formal and procedural perspective has frequently led to a disconnect between the public and policymakers, where the needs of the "silent citizens"—the large group of people in a country or group who do not express their opinions publicly—are buried under the voices of the "activists" minority of citizens. The traditional approach leaves critical insights unexamined, reducing civic engagement to tokenistic—symbolic rather than meaningful—gestures rather than facilitating authentic dialogue about what really matters to the majority of citizens as they go about daily life.

This chapter focuses on demonstrating how design thinking approaches and technology-mediated tools can transform civic engagement from procedural formalities into meaningful participation that captures the authentic needs of all citizens, especially the "silent majority" who typically remain disengaged from policy processes. Specifically, this chapter addresses the research question: How can citizen-centred design interventions overcome existing barriers to civic participation, enabling policymakers to access the lived experiences and genuine needs of citizens who traditionally do not engage in policy development?

By adopting citizen-centred approaches, we can transform civic participation in policymaking into a substantive process of problem-definition and solution-generation, so it becomes the core activity that it deserves, driving the development of effective policies that fulfil the real needs of ordinary citizens. This chapter argues for a transformative thinking where citizen engagement becomes an integral part of the problem-solving process, helping previously unengaged citizens to articulate their challenges and contribute to developing crafted policy solutions. The citizen-centred approaches based on design thinking can meticulously unpack the psychological barriers that inhibit civic participation, probing deep into the complex motivations and systemic

DOI: 10.4324/9781003667872-9

obstacles that silence citizens' voices. The citizen-centred approaches can be deployed to understand underlying factors of civic disengagement, leading to designing nuanced, context-specific civic engagement that addresses the root causes of citizens' reluctance and empowers meaningful policy participation.

Technology-mediated engagement methods such as digital platforms integrated with user-centred design principles can create adaptive, empathetic interfaces that lower psychological barriers to participation and amplify marginalized voices.

This chapter explores how design thinking can transform civic engagement by creating meaningful pathways for silent citizens to contribute authentically to policy development. We begin with a literature review that reveals a critical paradox: while citizens may formally "participate," they often remain genuinely unengaged, leading to policies that fail to reflect lived experiences. The literature review reveals three critical gaps that provide the rationale for this chapter's empirical studies:

First, while design thinking approaches show promise, there is limited empirical evidence of their effectiveness in real-world policy contexts, particularly in creating genuinely equal participation among diverse stakeholder groups. The *Policy Puzzle Game* study addresses this gap by empirically testing whether gamified co-design can overcome traditional power imbalances between experts and non-experts, activists and ordinary citizens in actual municipal policy development.

Second, existing digital engagement platforms largely fail to engage silent citizens, with most innovations benefiting already-active participants. The *Local Hero Nurturing Game* investigates whether combining gamification with psychological safety mechanisms can lower barriers sufficiently to engage non-activist citizens in meaningful policy contribution.

Third, current engagement methods require citizens to step outside their daily routines and adopt unfamiliar civic behaviours. The *24-Hour Project* explores whether leveraging natural social media behaviours can surface authentic policy needs without requiring conscious civic participation, potentially accessing the silent majority's perspectives through their everyday digital traces.

These studies collectively address the fundamental challenge of contemporary civic engagement: moving beyond procedural legitimacy toward substantive inclusion that captures the authentic needs and lived experiences of all citizens, particularly those systematically excluded from policy discourse. By examining different design intervention strategies—structured co-design, digital gamification, and behavioural leveraging—this chapter contributes to civic engagement that is both democratically legitimate and practically effective.

The chapter concludes by synthesizing these findings into actionable design strategies for engaging previously excluded voices and developing policies that genuinely reflect all citizens' needs and experiences.

Literature Review

The Evolution of Civic Engagement: From Democratic Imperative to Design Challenge

Scholarly discourse has predominantly framed citizen participation as a democratic imperative—essential for legitimacy, civic education, and political efficacy (Erickson & Nosanchuk, 1990; Cohen et al., 1972). Yet this procedural paradigm often reduces participation to what Arnstein (2020) termed "tokenism"—symbolic engagement after key decisions have been made, failing to redistribute power or influence meaningful outcomes (Figure 7.1).

Figure 7.1 Where it starts to go wrong.

Contemporary scholarship challenges this proceduralism. Verba et al. (1995) revealed how structural inequalities systematically exclude populations from participation. Likewise, OECD (2020) demonstrates traditional methods consistently engage only educated, resource-rich minorities. Majority opinions remain unexpressed where participation demands significant social, cultural, and economic capital (Dalton, 2018; Schlozman et al., 2012). This gave rise to the concept of "silent majority" (Nixon, 1969). In this research, we refer to the citizens who remain unengaged in the policy process as "silent citizens".

The disconnect between procedural participation and genuine engagement manifests through systematic exclusion and structural limitations. King et al. (1989) documented how policy discussions exclude those lacking technical expertise, while den Besten et al. (2008) found that participation environments feel "unfamiliar and uncomfortable" for newcomers. Kim and Nam (2017) demonstrate that even well-intentioned participatory workshops devolve into "procedural formalities" where citizens endorse predetermined decisions rather than co-create solutions. Despite innovations like participatory budgeting, Lee (2019) argues success is measured through participant numbers rather than substantive policy impact.

The common formats—public hearings, roundtables, and co-design sessions—tend to alienate non-expert participants, reproducing dominance by experienced actors and discouraging equal voice (den Besten et al., 2008; Kim & Nam, 2017; Kim et al., 2019). States often equate attendance with consent, reducing citizen roles to approval rather than problem co-definition. The result is a recurrent paradox: citizens "participate," yet remain unengaged; ideas are generated, yet rarely translated into policy; and the daily realities of the "silent citizens" stay off the policy radar (Figure 7.2).

Figure 7.2 Unseen below the beam: the silent citizens beyond the policy radar.

The Challenge of Silent Citizens: Beyond Statistical Representation

The literature reveals a critical distinction between achieving statistical representation and gaining access to authentic citizen voices. While demographic representation remains important, scholars increasingly recognize that the "silent citizens"—those who choose not to participate in formal policy processes—often hold crucial insights about policy impacts and unmet needs (Kim et al., 2018b). France's Yellow Vests protests and South Korea's minimum wage failures demonstrate how even well-intentioned policies can backfire when ignoring citizens' lived experiences. France's climate-justified fuel tax overlooked rural commuting needs and stagnant wages, while South Korea's wealth-redistribution efforts outpaced small-business capacity and regional/sectoral realities (Grossman, 2019; Yim & Kim, 2023). Both cases demonstrate that policies can cause widespread backlash and distrust without meaningful citizen engagement.

This challenge is compounded by what Crivellaro et al. (2014) identify as the "activist bias" in civic engagement—the tendency for participation mechanisms to amplify already-vocal minorities while further marginalizing quiet majorities. The paradox, as Kim et al. (2019) articulate, is that policies must serve everyone, including non-participants, yet policymakers inevitably rely on those willing to engage, creating a feedback loop that systematically excludes certain perspectives and experiences.

Listening to silent citizens is different from simply gathering majority opinions or achieving statistical representation. It means extending reach beyond activist citizens to include typically uninvolved individuals who may offer valuable perspectives from personal experience when properly facilitated. Unlike activists who tend to highlight topical issues, silent citizens often surface everyday concerns that, though seemingly private or trivial, may reflect broader societal problems worthy of public attention.

How, then, can we engage the silent citizens? The solution requires examining why citizens are reluctant to participate and designing features that address these barriers. We need methods that not only accommodate citizens who want to speak up, but also encourage silent citizens to want to express themselves. This seems to be a difficult task, and it actually is. However, achieving this goal becomes possible through design thinking approaches that enable nuanced, systematic understanding of citizens' needs and motivations (Kimbell, 2015; Fischer, 2009).

Design Thinking as a Game Changer in Policy Engagement

Research demonstrates that design-led methods effectively uncover participation barriers for less vocal citizens while enabling the creation of more inclusive, responsive, and effective civic engagement strategies (Kimbell, 2015; Bason, 2016; OECD, 2020; Bovaird & Loffler, 2013; Light & Akama, 2012). By deeply understanding people's participation needs and designing citizen-centred methodologies, we can bridge the gap between policymaking and the silent citizens whom it ultimately serves.

A growing body of literature advocates for design thinking approaches as a transformative alternative to traditional civic engagement methods. Kimbell (2015) argues that design thinking's user-centred (or citizen-centred in the policy context) methodologies can fundamentally reorient policy development from expert-driven processes to citizen-centred problem-solving. This perspective aligns with Bason's (2016) framework for "design for policy," which positions citizens not as subjects of policy but as active partners in co-defining problems and co-creating solutions.

The theoretical foundation for this shift draws from Cohen et al. (1972) "garbage can model," which recognizes policymaking as a complex, non-linear process where problems, solutions, and participants interact in unpredictable ways, much like how garbage is thrown into a can. Design thinking approaches acknowledge this complexity, understanding design problems as "wicked problems" (Buchanan, 1992) akin to policy problems (Head, 2022), while providing structured methods for navigating it. Fischer (2009) argues that such approaches can bridge the gap between technical expertise and democratic participation by creating accessible frameworks for citizen contribution. Further, empirical evidence supports the efficacy of design-led approaches. Light and Akama (2012) demonstrate how facilitated co-design processes can empower communities to articulate needs that traditional consultation methods fail to capture.

Bovaird and Loffler (2013) document that co-creation methods positioning citizens as partners rather than consumers lead to more effective public services. However, significant gaps remain in understanding how to implement these approaches across diverse policy contexts, particularly in engaging those who typically choose not to participate.

Technology-Mediated Participation: Promise and Limitations

Among design thinking-based citizen participation methods, technology-mediated approaches are gaining significant attention. In particular, the Human-Computer Interaction (HCI) field has been actively exploring various methodologies to enhance citizen engagement. In the context of civic engagement, Kim et al. (2018a) define design intervention as a creative design solution with the purpose of inducing citizens' behavioural changes into a more matured state for policy participation. They introduced several such cases including media facades that visualise the text of passersby's street speech (Fortin et al., 2014) and interactive public poll visualisation via wall projection (Valkanova et al., 2014) (Figure 7.3). Similarly, Kim et al. (2021) developed an infographic website to visualise public opinion polls on controversial topics.

However, critical assessments reveal that many digital platforms merely transfer analogue participation methods to digital formats without addressing fundamental motivational and psychological barriers (Elsen & Schelings, 2019; Van Stolk et al., 2017). These platforms

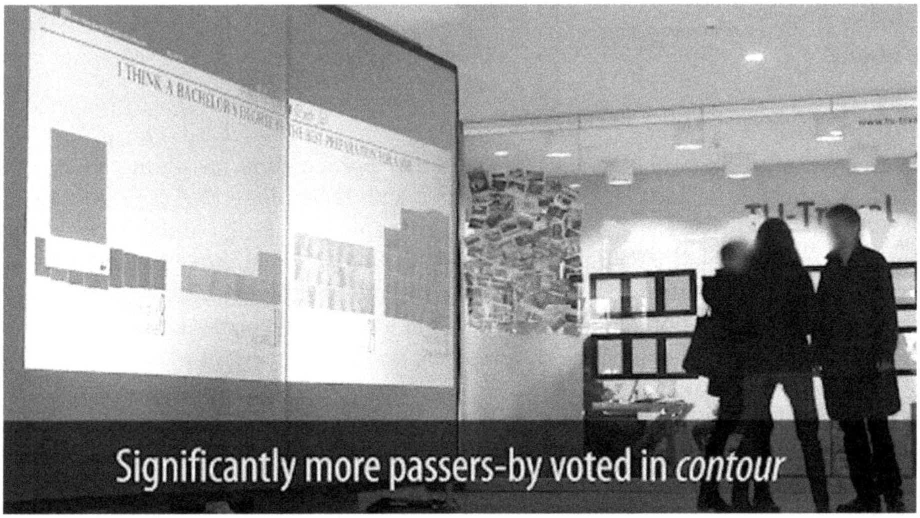

Figure 7.3 MyPosition (Valkanova et al., 2014): A camera-embedded wall projection system that detects passersby's hand-raise gestures and visualises real-time polling results.

retain the traditional problem of passive, 'tokenistic' participation discussed previously, where users merely respond to predetermined policy agenda. Jasim et al. (2021) found that even sophisticated platforms like CommunityClick struggle to engage beyond typical activist users. Kale et al. (2024) demonstrate through their analysis of participatory budgeting in Chicago that technical sophistication alone cannot overcome the social and psychological barriers to participation. Technology is not a magic wand in itself. Simply enhancing accessibility or convenience without tackling the core reasons why people avoid voting or hesitate to express their views publicly will not effectively promote broad public participation. As with the traditional, non-technological civic engagement, the major design challenge in developing technology-mediated civic engagement is to find effective ways to engage citizens and listen to their unique voices (Berigüete et al., 2024; Elsen & Schelings, 2019; Van Stolk et al., 2017).

Gamification and Behavioural Design: Emerging Frontiers

In developing design thinking-based engagement methods for silent citizens, gamification can serve as a powerful tool. An emerging strand of literature explores gamification as a strategy for overcoming participation barriers. Breuer et al. (2022) argue that game mechanics can transform civic engagement from "duty-driven" to "desire-driven" participation. This aligns with behavioural design principles that leverage intrinsic motivations rather than extrinsic pressures to encourage engagement. The theoretical foundation for gamification in civic contexts draws from self-determination theory, which posits that autonomy, competence, and relatedness drives intrinsic motivation (Hassan, 2017). Huang et al. (2015) demonstrate how gamified elements can transform passive users into active contributors, though questions remain about the depth and quality of engagement achieved through such methods. The challenge, as identified in current literature, is moving beyond superficial gamification—"pointsification" offering quick, extrinsic-rewards such as points, badges and leaderboard—toward meaningful game design that cultivate genuine engagement, reflection, and civic contribution. Therefore, the three empirical studies presented in this chapter adopt gamification as a key driver to develop novel citizen engagement strategies and to apply them empirically.

Empirical Studies

Participated, But Not Engaged: Policy Puzzle Game

In order to tackle the problem of non-engagement while participating, as discussed in 7.2, Designize Lab at Korea Advanced Institute of Science and Technology (KAIST) developed "Policy Puzzle Game" (Kim & Nam, 2022). The research aimed to design a policy co-design workshop process for every participant to be fully engaged on an equal ground without being alienated, and to empirically evaluate the effect of the process on participant engagement, policy idea generation, and the realization of co-design values in a real-world context.

The research methodology included the development and deployment of a gamified policy co-design workshop—the "Policy Puzzle Game"—in a municipal youth policy workshop in Yuseong, Daejeon, South Korea. The workshop involved 30 participants comprising ordinary citizens, youth activists, civil servants, and design thinking facilitators. Diverse stakeholder groups collaborated within a structured, game-based environment where ideas and constraints were explored using jigsaw puzzle-style toolkits (Figure 7.4). Data were collected through semi-structured interviews and onsite observations, and then analysed thematically with intercoder reliability checks.

Figure 7.4 Policy puzzle game workshop.

The game was crafted as a serious game to transform policy co-design into an engaging, rule-based collaborative experience designed to ensure both feasibility and acceptability of citizen-generated policy ideas. The game employs five types of hexagonal puzzle pieces—Topic, Idea, Constraint, Local Information, and Existing Policy—that participants place on a honeycomb board, with the visual tessellation making discussion routes, consensus-building, and trade-offs tangible and transparent. Game rules establish procedural and substantive equality by requiring mixed groups of citizens, civil servants, and facilitators (design researchers); distributing equal numbers of Idea and Constraint pieces to each participant; and mandating unanimous consent before any piece can be placed on the board, effectively preventing domination while encouraging reasoned persuasion.

The study found that the visual and board game style toolkit fostered vibrant social interaction, making complex idea negotiation both transparent and accessible. This approach helped participants with varying levels of policy expertise and understanding contribute meaningfully while learning from one another. The game design successfully transcended the hierarchical barriers that often emerge in co-design processes when expert and non-expert participants work together in mixed groups. The honeycomb board served as a powerful "boundary object" (Leigh Star, 2010) that externalized reasoning paths and decision-making routes for people with different backgrounds and varying levels of understanding, making every participant's contributions and the communication process itself highly visible—a transparency that encouraged equal engagement and prevented domination by more vocal stakeholders. The Policy Puzzle Game was found to facilitate equal and meaningful participation among all groups, transforming traditional power dynamics: (1) ordinary citizens were empowered to voice their needs and provide fundamental policy rationale based on their lived experiences as civic input. (2) The activists discovered a new collaborative role in localizing abstract policy concepts into familiar neighbourhood action items and bridging the gap between citizens and policy processes; and (3) civil servants shifted from being sceptical observers or evaluators to active co-players who contributed practical constraints that enhanced feasibility rather than simply blocking ideas.

Crucially, the game rules—particularly the requirement for unanimous consent before placing any puzzle piece—ensured that no participant could dominate discussions, while the equal distribution of the puzzle pieces provided the participants with procedural equality. Civil servants specifically appreciated that the game helped them overcome their usual scepticism toward citizen participation by providing a novel way of capitalizing on citizens' civic input while contributing their own policy implementation knowledge, ultimately leading to more realistic and implementable policy solutions that satisfied both citizen needs and administrative feasibility requirements.

The research's key contribution lies in demonstrating how combining design thinking and gamification, which ensured fair and equal policy participation can overcome longstanding barriers to effective citizen participation in policymaking, enabling different policy stakeholders to fully demonstrate their own civic expertise. The game's unique approach enabled ordinary citizens to become genuine protagonists in policy formation by ensuring balanced interaction between activists and non-activist citizens, as well as experts and non-experts, creating a structured environment where participants can collectively develop and refine policy ideas without alienation, making them more feasible and acceptable. By gamifying the participatory policy development process, it breaks down traditional barriers between civil servants and citizens, enabling more inclusive and creative policymaking. Ultimately, the approach contributes to transforming participatory exercises from mere procedural formalities to genuinely valuable, implementation-ready idea generation.

***Overcoming Barriers to Participation:* Local Hero Nurturing Game**

As addressed in the 7.2 literature review, existing digital platforms have not been able to overcome the limitations of traditional methods where participation is predominantly driven by activists. To overcome these limitations, KAIST Designize Lab applied a design thinking—citizen-centred—approach that digs deeper into the real needs and motivations that shape how people engage—or don't engage—in policymaking on digital platforms. The study aimed to design a novel digital engagement platform targeting non-activist citizens and examine the effectiveness of the platform in lowering participation barriers. Based on such a design thinking approach, different from the usual top-down approaches, existing theories were analysed to understand the psychological and emotional barriers citizens face in policy development (e.g. the fear of exposing personal matters in public and being seen as a whistle-blower). As a result, a gamification strategy was created to address these barriers. Then, applying gamification theory—proven effective in encouraging participation—built a digital platform that uses gameplay to engage citizens.

The platform, called the '*Local Hero Nurturing Game,' is a* web-based collaborative game where participants define local issues and propose solution ideas via public kiosks and smartphones.

The research methodology included designing, prototyping, and field deployment of the game at the National Failure Expo, hosted by South Korea's Ministry of the Interior and Safety. The field study was conducted at the Expo over a three-day period. Data were collected through field observation, walk-in interviews, and user-generated content on the platform. The participants generated 247 pieces of content, and the observation and interview data were collected onsite. These data were analysed by directed content analysis to identify how the platform engaged citizens by different features of the game. Follow-up interviews with local civil servants were also conducted to evaluate the platform's efficacy for useful policy idea generation.

The game aims to transform civic engagement into an immersive gaming experience where players nurture a young, genderless hero character who needs their help to solve local community

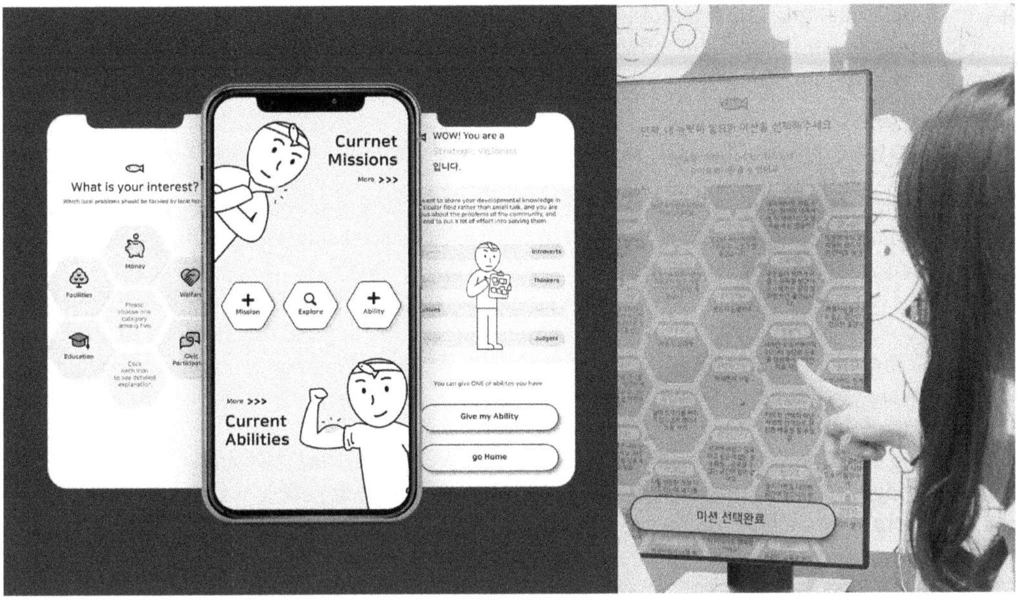

Figure 7.5 Local hero nurturing game (left: game pages, right: honeycomb interface).

Source: Designize Lab., KAIST.

problems. The game lets the players, as local residents, assign "missions" and "solutions" (real neighbourhood issues and solution ideas) to their digital hero empowered by the players' own abilities. Players create hexagonal tiles on a dynamic honeycomb board, linking missions and solutions through intuitive relationships (Figure 7.5). The game's unique contribution lies in making civic participation feel protective and altruistic: residents naturally propose specific, feasible solutions because they want to help their vulnerable hero succeed. Through this playful, collective problem-solving and collaborative ideation, ordinary citizens can find civic engagement pleasurable and fear-free.

The research found that the game transformed civic participation into an interactive and empowering experience. At its core was the game character, a personified local hero who relies on residents' guidance to tackle community challenges, encouraging collaborative exploration and definition of local issues. The participants commented that the game enabled them to visualize and organize issues collectively, enhancing a sense of ownership for the local community and shared purpose. By presenting the hero as an approachable figure in need of support, the game appeared to foster empathy and intrinsic motivation among participants, particularly those with no prior experience in civic engagement.

The gamification strategy on a digital platform used in the research tackled the psychological hurdles that kept people quiet. The platform reduced the burden of exposing individual problems directly by providing an in-game avatar through whom the participants could project personal issues. The fear of being perceived as a whistle-blower was alleviated by reframing problem-raising as enjoyable and constructive mission assignments within the game environment, transforming potentially confrontational civic engagement into playful and healthy community participation. The local officials appreciated that the game helped turn passive residents into active contributors and gave the officials real insight into community needs. The research demonstrated that the game transformed hesitation into involvement, creating a psychologically safe and empowering experience that validates citizens' perspectives, revitalizing democracy through majority involvement.

Tapping Into What They Do Anyway: 24-Hour Project

The third empirical study aims to tackle the participation barrier of apathy—a typical attitude of the "silent citizens"—and to develop a design thinking-based, technology-mediated approach to engage many citizens indifferent to civic engagement.

Deeply examining citizens' needs can elicit behavioural change in people, because relying solely on conscience to drive behavioural change is often inadequate for influencing most individuals. Rather, desire can be a powerful driver for behavioural change—tapping into what they actually "want" to do instead of what they "should" do. This intrinsic yet frequently overlooked force forms a foundational aspect of the design thinking approach.

Adopting this approach, Designize Lab at KAIST launched the "24-Hour Project" to take online sharing on a social media platform—a kind of activity that people want to do every day—and transform it into useful policy insights (Kim et al., 2018b). By encouraging young users to post photos and tag activities and locations on Facebook, the initiative leveraged their natural desire for self-expression to collect authentic data (Figures 7.6 and 7.7).

This study aimed to examine the potential of casual social media interactions to identify latent policy needs. A three-month Facebook intervention targeting youth in Daejeon, South Korea, generated 330 posts with associated engagement metadata. Unstructured data, including qualitative comments, were subjected to content analysis and coded by six researchers, with

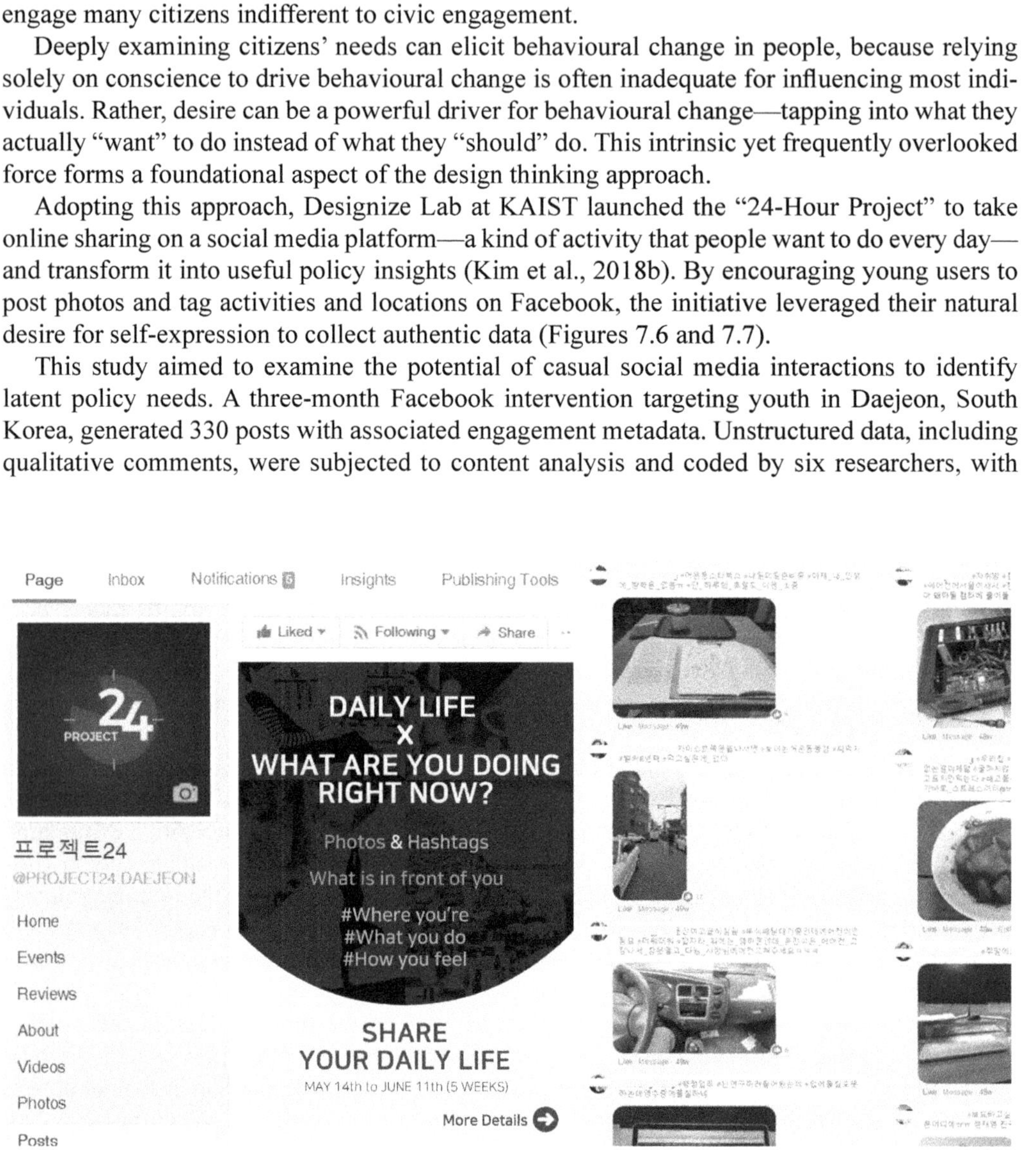

Figure 7.6 Facebook page to collect comments about citizens' daily lives.

Source: Designize Lab., KAIST.

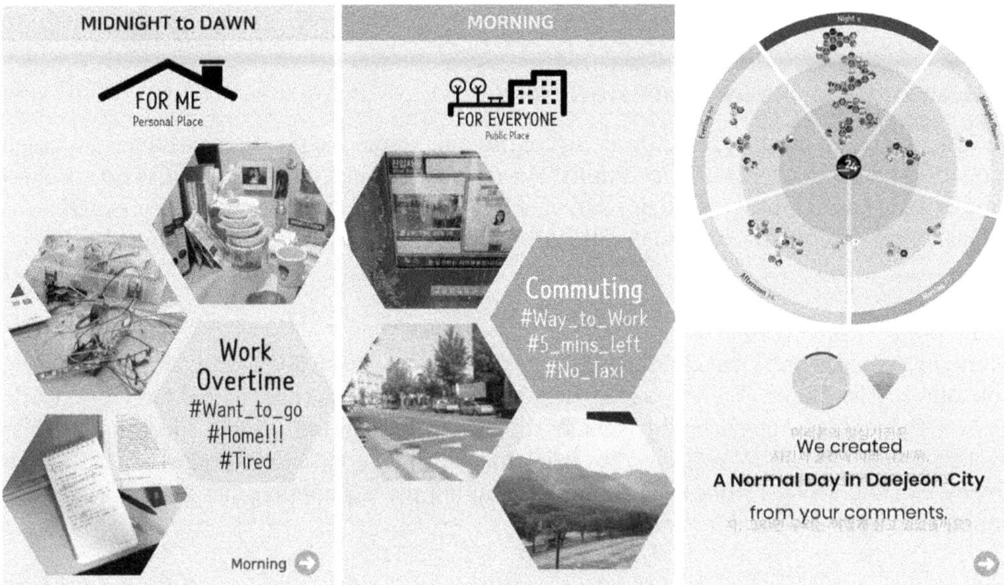

Figure 7.7 A Facebook intervention sharing daily lives through citizen comments to spark dialogue.

Source: Designize Lab., KAIST.

high intercoder reliability. Engagement patterns were quantitatively analysed across three iterative phases of civic participation. Findings were subsequently evaluated with local civil servants and youth groups to determine policy relevance.

The Facebook intervention comprised three iterations. The participants began responding to the simple question, "What are you doing now?" In the first iteration, participants shared playful complaints. It was found that this type of playful, game-like communication culture facilitated the expression of everyday concerns among citizens. In the second iteration, individual complaints were reframed as collective issues pertinent to specific policy target groups, such as "job seekers' concerns." The research team analysed these group-level expressions, based on which easily relatable personas were developed, which served as stimuli for the second iteration. In the third iteration, citizens treated these personas as declarations of their own needs, sharing them actively through their personal accounts and contributing comments that were notably more serious and in-depth than before.

Now, what started as casual complaints quickly revealed systemic problems faced by the policy target groups. A post about night shifts exposed broader labour issues, while gripes about local services highlighted real community needs. By tapping into people's natural urge to share and complain online, the project captured authentic voices that formal policy channels never reach—the "silent citizens" who avoid civic meetings but freely vent online.

The result was a rich trove of real stories in which everyday frustrations became valuable policy resources. The civil servants involved in the project said that it helped the local government develop more effective youth policies by revealing the real needs of the policy target groups. The approach made civic participation effortless by engaging ordinary citizens in normal social media activity, seamlessly gathering public sentiment, breaking down the barriers of indifference to civic participation, and delivering policies grounded in the lived experiences of the community.

Discussions: Beyond Procedural Participation

The chapter makes meaningful knowledge contributions from the three perspectives presented below by addressing the research question, "How can citizen-centred design interventions overcome existing barriers to civic participation, enabling policymakers to access the lived experiences and genuine needs of citizens who traditionally do not engage in policy development?"

Transforming Participation from Duty to Desire

Our findings demonstrate that civic engagement transforms when it aligns with citizens' intrinsic motivations. The *Policy Puzzle Game* created procedural fairness enabling hidden voice, while the *Local Hero Game* engineered psychological safety through avatars that addressed the fear of participation. The *24-Hour Project* revealed that authentic civic input emerges naturally from already existing social media behaviours, challenging the assumption that civic engagement requires conscious political motivation.

Expanding the Civic Engagement Paradigm Toward Implementable Civic Innovation

Our research contributes a new approach to inclusive civic design that adapts participation structures to citizen needs rather than expecting citizens to adapt to existing formats. This challenges the "Tokenism" and encourages a move toward "Citizen Control" in Arnstein's (2020) participation ladder by demonstrating that different citizens may require entirely different pathways to meaningful engagement. Implementation of our three design strategies requires careful attention to context. Equitable engagement design must establish clear rules preventing domination—our unanimous consent requirement proved essential and can be made efficient and even fun through creative design of a "serious game". Effective civic gamification requires understanding specific participation barriers rather than superficial entertainment. Leveraging natural behaviours demands sophisticated understanding of citizens' existing daily practices, as demonstrated by transforming social media complaints into policy intelligence.

Transforming Democratic Practice

Our contribution extends beyond methodology to transformative thinking: (1) challenging the "rational citizen" assumption (Kuklinski & Quirk, 2000) by showing citizens already possess policy-relevant insights through lived experiences without needing political knowledge; (2) redefining democratic legitimacy (Erickson & Nosanchuk, 1990) as authentic inclusion of diverse experiences rather than just formal procedures; (3) expanding participation theory beyond Arnstein's (2020) hierarchical progression to multiple parallel pathways, as well as quantum leaps from non-participation to meaningful and substantive participation, leading to citizen power; and (4) reconceptualizing citizens as distributed experts rather than subjects needing civic education. Our research demonstrates that silent citizens' silence may reflect design failures in civic institutions rather than citizen apathy. The transformation from procedural to substantive participation requires recognizing that different citizens contribute most effectively through different mechanisms, suggesting democratic systems must develop diverse pathways while maintaining commitment to equality and inclusion.

Summary and Conclusions

The new way of looking at civic engagement advocated in this chapter does not simply mean fulfilling the 'procedural requirements' for democratically developing policy solutions by meeting only the nominal conditions of 'participation'. The chapter examined how traditional civic engagement methods often reduce citizen participation to mere procedural formalities and current digital platforms limit citizens to reactive roles after key decisions have been made. In response, it argued for the advantages of citizen-centred design thinking approaches. These approaches enable more powerful and genuine problem-solving, making it possible to uncover the rich, vivid, and nuanced stories that only the problem holders can tell. Next, the chapter demonstrated how gamified interventions and technology-mediated platforms could break down participation barriers by designing inclusive, accessible environments that encouraged equal contribution from diverse stakeholders. Finally, the chapter revealed how tapping into citizens' natural behaviours—such as social media sharing and everyday complaints—could transform casual interactions into valuable policy insights, enabling silent citizens to contribute authentically without the traditional burdens of policy participation.

Derived from our research findings are new and creative design strategies for civic engagement:

1. Design Equitable Engagement
   Establish clear rules to ensure all citizens can contribute their expertise on equal terms.
2. Adopt Gamification Techniques
   Incorporate engaging, gamified elements to make participation psychologically safe, activities socially enjoyable, and quests easily achievable.
3. Tap into Natural Daily Behaviours
   Leverage ordinary actions and motivations from daily life as entry points for authentic civic input.

The ultimate goal of civic engagement is, then, to gain a broad understanding of the needs and challenges that citizens genuinely face—whether widely known or not—and to refine and define them as authentic policy issues through design thinking approaches, thereby developing policies that are practically effective from the perspective of the citizen beneficiaries by designing engagement for the majority of previously disengaged citizens.

References

Arnstein, S. (2020). "A ladder of citizen participation": Journal of the American Institute of Planners (1969). In R. T. LeGates, F. Stout & R. W. Caves (Eds.), *The City Reader* (pp. 290–302). Routledge. https://doi.org/10.4324/9780429261732

Bason, C. (2016). *Design for Policy*. Routledge.

Berigüete, F. E., Santos, J. S., & Cantalapiedra, I. R. (2024). Digital revolution: Emerging echnologies for enhancing citizen engagement in urban and environmental management. *Land*, *13*(11), 1921. https://doi.org/10.3390/land13111921

Bovaird, T., & Löffler, E. (2013). From engagement to co-production: How users and communities contribute to public services. In V. Pestoff, T. Brandsen, & B. Verschuere (Eds.), *New Public Governance, the Third Sector, and Co-production* (pp. 35–60). Routledge. https://doi.org/10.4324/9780203152294

Breuer, H., Bessant, J., & Gudiksen, S. (2022). Chapter 1: Challenges: Gamification and games to tackle grand challenges. In *Gamification for Innovators and Entrepreneurs: Using Games to Drive Innovation and Facilitate Learning* (pp. 16–30). De Gruyter. https://doi.org/10.1515/9783110725582-005

Buchanan, R. (1992). Wicked problems in design thinking. *Design Issues*, *8*(2), 5–21. www.jstor.org/stable/1511637

Cohen, M. D., March, J. G., & Olsen, J. P. (1972). A garbage can model of organizational choice. *Administrative Science Quarterly*, *17*(1), 1–25.

Crivellaro, C., Comber, R., Bowers, J., Wright, P. C., & Olivier, P. (2014, April). A pool of dreams: Facebook, politics and the emergence of a social movement. In *Proceedings of the SIGCHI Conference on Human Factors in Computing Systems* (pp. 3573–3582). ACM.

Dalton, R. J. (2018). *Citizen Politics: Public Opinion and Political Parties in Advanced Industrial Democracies*. CQ Press.

den Besten, O., Horton, J., & Kraftl, P. (2008). Pupil involvement in school (re) design: Participation in policy and practice. *Co-Design*, *4*(4), 197–210.

Elsen, C., & Schelings, C. (2019). Citizen participation through digital platforms: The challenging question of data processing for cities. In *Proceedings of SMART 2019: The Eighth International Conference on Smart Cities, Systems, Devices and Technologies* (pp. 19–25). International Academy, Research, and Industry Association (IARIA).

Erickson, B. H., & Nosanchuk, T. A. (1990). How an apolitical association politicizes. *Canadian Review of Sociology/Revue canadienne de sociologie*, *27*(2), 206–219.

Fischer, F. (2009). *Democracy and Expertise: Reorienting Policy Inquiry*. OUP Oxford.

Fortin, C., Neustaedter, C., & Hennessy, K. (2014, June). The appropriation of a digital "speakers" corner: Lessons learned from the deployment of mégaphone. In *Proceedings of the 2014 Conference on Designing Interactive Systems* (pp. 955–964). ACM.

Grossman, E. (2019). France's yellow vests–symptom of a chronic disease. *Political Insight*, *10*(1), 30–34.

Hassan, L. (2017). Governments should play games: Towards a framework for the gamification of civic engagement platforms. *Simulation & Gaming*, *48*(2), 249–267.

Head, B. W. (2022). *Wicked Problems in Public Policy: Understanding and Responding to Complex Challenges* (p. 176). Springer Nature.

Huang, S.-W., Suh, M. M., Hill, B. M., & Hsieh, G. (2015). How activists are both born and made: An analysis of users on change. org. In *Proceedings of the 33rd Annual ACM Conference on Human Factors in Computing Systems* (pp. 211–220). ACM.

Jasim, M., Khaloo, P., Wadhwa, S., Zhang, A. X., Sarvghad, A., & Mahyar, N. (2021). CommunityClick: Capturing and reporting community feedback from town halls to improve inclusivity. *Proceedings of the ACM on Human-Computer Interaction*, *4*(CSCW3), 1–32.

Kale, A., Liu, D., Ayala, M. G., Schwab, H., & McNutt, A. (2024). What can interactive visualization do for participatory budgeting in Chicago? *IEEE Transactions on Visualization and Computer Graphics*, *31*(1), 415–425.

Kim, C., Kim, J., & Nam, K. Y. (2019). Beyond a workshop: New design opportunities for participatory policy development. *Design Works*, *2*(1), 12–24.

Kim, C., & Nam, K. Y. (2017). Policymaking and the design workshop: A new business direction. *Design Management Review*, *28*(4), 14–20.

Kim, C., & Nam, K. Y. (2022). Policy puzzle game: Making policy ideas feasible and acceptable in policy co-design. *CoDesign*, *18*(4), 448–465.

Kim, C., Woo, E., & Nam, K. Y. (2018a). Designing engagement for local public policy development. In *Proceeding of the 2018 dmi: Academic Design Management Conference*. Design Management Institute (DMI). *Refer to this article for bibliographical information of individual cases.

Kim, C., Woo, E., & Nam, K. Y. (2019). The paradox of the excluded majority in youth policy design. *Design Management Review*, *30*(3), 14–21.

Kim, C., Yang, H., Jang, S., & Nam, K. Y. (2018b, June). Grumble to policy need: Deriving public policy needs from daily life on social media platform. In *Proceedings of the 2018 Designing Interactive Systems Conference* (pp. 971–983). ACM.

Kim, H., Kim, H., Jo, K. J., & Kim, J. (2021). StarryThoughts: Facilitating diverse opinion exploration on social issues. *Proceedings of the ACM on Human-Computer Interaction*, *5*(CSCW1), 1–29.

Kim, K. M. (2023). Why did the moon Jae-in administration's minimum wage policy retreat? ? *Journal of Korean Politics*, *32*(2), 31–64. https://iris.uniroma1.it/handle/11573/1730093
Kimbell, L. (2015). *Applying Design Approaches to Policy Making: Discovering Policy Lab*. University of Brighton.
King, S., Conley, M., Latimer, B., & Ferrari, D. (1989). *Co-Design: A Process of Design Participation*. Van Nostrand Reinhold.
Kuklinski, J. H., & Quirk, P. J. (2000). Reconsidering the rational public: Cognition, heuristics, and mass opinion. In A. Lupia, M. D. McCubbins, & S. L. Popkin (Eds.), *Elements of Reason: Cognition, Choice, and the Bounds of Rationality* (pp. 153–182). Cambridge University Press.
Lee, D. (2019). Citizen participation system: Is there a citizen? In *Proceedings of the Seoul Association for Public Administration Conference* (pp. 19–37). Seoul Association for Public Administration (SAPA).
Leigh Star, S. (2010). This is not a boundary object: Reflections on the origin of a concept. *Science, Technology, & Human Values*, *35*(5), 601–617.
Light, A., & Akama, Y. (2012, August). The human touch: Participatory practice and the role of facilitation in designing with communities. In *Proceedings of the 12th Participatory Design Conference: Research Papers-Volume 1* (pp. 61–70). ACM.
Nixon, R. M. (1969, November 3). *Address to the Nation on the War in Vietnam [Transcript]*. The American Presidency Project, University of California. www.presidency.ucsb.edu/documents/address-the-nation-the-war-vietnam
OECD. (2020). *Innovative Citizen Participation and New Democratic Institutions: Catching the Deliberative Wave*. OECD Publishing. https://doi.org/10.1787/339306da-en.
Schlozman, K. L., Verba, S., & Brady, H. E. (2012). *The Unheavenly Chorus: Unequal Political Voice and the Broken Promise of American Democracy*. Princeton University Press.
Valkanova, N., Walter, R., Vande Moere, A., & Müller, J. (2014, February). MyPosition: Sparking civic discourse by a public interactive poll visualization. In *Proceedings of the 17th ACM Conference on Computer Supported Cooperative Work & Social Computing* (pp. 1323–1332). ACM.
Van Stolk, C., Manville, C., Devaux, A., & Dubow, T. (2017). *Civic Engagement: How Can Digital Technologies Underpin Citizen-Powered Democracy?* RAND Corporation. Retrieved from https://coilink.org/20.500.12592/0bkfj0
Verba, S., Schlozman, K. L., & Brady, H. E. (1995). *Voice and Equality: Civic Voluntarism in American Politics*. Harvard University Press.

Section III

Holistic Sustainability and Way of Living

8 From Design to Adoption

A Responsible Approach to Stakeholder Participation in Digital Health Innovations

Hong Li and Yangyang Zhao

Introduction

Digital technologies are increasingly recognized as critical drivers in shaping the future of global health, with the World Health Organization (WHO) emphasizing their role in strengthening health systems through its *Global Strategy on Digital Health 2020–2025*. The strategy aims to strengthen health systems by leveraging digital innovations for consumers, health professionals, healthcare providers, and industry, with the ultimate goal of empowering patients and achieving the vision of "health for all" (WHO, 2021). Achieving such an ambitious goal requires broad stakeholder engagement, as the challenges associated with operational responsibilities are intertwined throughout the entire process, from conceptualization to the adoption of digital health innovations. The utmost importance of responsibility is highlighted as a core priority in digital health, aligning with the principles of responsible (research and) innovation (hereafter RI), which seeks to benefit individuals and communities in an ethically acceptable, sustainable, and socially desirable manner (von Schomberg, 2013; WHO, 2021).

Although responsible development has gained traction in diverse domains such as digital remote care, the predominant focus has remained on the innovation process itself (Stilgoe et al., 2020; Pavie et al., 2014; Lehoux et al., 2018). The adoption phase, occurring after innovation, has received comparatively little attention, despite its crucial role in ensuring responsible practices. In the healthcare sector, responsible adoption is particularly important due to the far-reaching impact of digital innovation on diverse stakeholders, including patients, health professionals, hospitals, and other involved parties. However, there is limited research in the digital health domain that thoroughly scrutinizes the design of responsible adoption across stakeholders. Such studies would provide deeper insights into the responsible adoption of digital health innovations.

Pioneering studies (Lynn, 1993; Kipnis, 1993; van Berkel & Bos, 1999) have long recognized the connection between technological innovation and design. Nonetheless, much of the design literature has traditionally concentrated on theoretical discourse rather than in-depth exegesis of design methods (Oxman, 2006). A shift away from conventional theoretical and methodological perspectives on design toward more context-driven research activities was emphasized by van Berkel and Bos (1999). This shift holds significance for developing systematic explanations of how design methods operate in practice and for generating insights into how digital innovations can be responsibly integrated into existing healthcare systems. In this context, we seek to advance the understanding of participatory design (PD) methodology and its application in fostering RI in digital remote care adoption.

This study contributes to the ongoing debate on the usability of PD in the digital age by extending its scope to encompass the post-innovation (i.e., adoption) stage and contextualizing its application in the responsible and digital innovation context. Through a real-world case study,

DOI: 10.4324/9781003667872-11

it explores stakeholder participation in the responsible adoption of digital health innovations for remote care delivery, providing actionable approaches and insights for design initiatives that promote responsible integration into healthcare systems.

Theoretical Foundation

Digital Remote Care and Its Responsible Adoption

The COVID-19 pandemic has accelerated the adoption of digital technologies for delivering a range of health innovations. One notable example is the emergence of digital remote care as an innovative approach to address growing demand and resource constraints within the healthcare sector. The definition of digital remote care extends beyond simply incorporating digital technologies into healthcare work practices. It facilitates the monitoring of patients' vital signs, mobility, and overall safety within their homes, while also placing strong emphasis on patient empowerment (Barlow et al., 2002). This typically involves training patients in relevant skills, enabling them to take an active and responsible role in managing or mitigating the impact of diseases on their physical health. Such empowerment is vital when patients utilize digital technologies to access care outside conventional healthcare settings.

In parallel, responsibility in research and practice has gained increasing attention in the realm of digital health innovations. This reflects a shared objective, that is, aligning technological innovations and adoptions with the needs and values of individuals and communities while minimizing potential negative impacts. The concept of RI has become a cornerstone of guiding principles for decision-making and practice at the organizational level (von Schomberg, 2013). RI underscores the importance of broad inclusion of individual social actors across both innovation development and organizational governance processes (Silva et al., 2019). We argue that RI should be expanded to encompass the entire innovation life cycle, spanning from the initial understanding of societal needs to practical realization through operationalization. For instance, Stilgoe et al. (2020) developed a framework for responsible innovation that emerged from various contexts. By further considering the societal actors' needs and benefits in responsible research and practice, the framework presented four key dimensions: "inclusion", "anticipation", "reflexivity", and "responsiveness" (Stilgoe et al., 2020). The "inclusion" dimension encourages broad and inclusive engagement with diverse stakeholders. The "anticipation" dimension involves analyzing both the intended and unintended impacts of innovation. The "reflexivity" dimension focuses on examining underlying motivations and potential impacts. Lastly, the "responsiveness" dimension leverages collective reflexivity to shape the direction and the trajectory of innovation through participatory governance.

Contextualizing Participatory Design (PD) for the Responsible Adoption

Several studies have begun to explore the role of design in advancing responsible research and practice. Examples include the use of design thinking to create responsible innovation (Pavie & Carthy, 2015), design strategies that address the social and environmental consequences of responsible innovation during production (Bentham, 2017), and a three-part process for engaging (co-)design in responsible innovation (Craigon et al., 2023). Yet, this growing body of work still largely follows traditional design models, typically structured around five sequential stages: empathize, define, ideate, prototype, and test. Such a design scope is not inclusive of the adoption of an innovation, leaving a critical gap in achieving responsibility in the post-innovation phase. Moreover, the prevailing emphasis on theoretical discourse has not yielded in-depth explanations of design methods within specific contexts (Oxman, 2006).

To bridge this gap, we embarked on an exploration of established design approaches that can facilitate stakeholder participation, contextualizing them for our specific setting of responsible adoption. We propose PD as a particularly relevant approach to support this process. Originating in Scandinavia in the 1970s and 1980s, PD was driven by the goal of empowering individuals to maintain control over the development and refinement of work practices. In today's healthcare landscape, PD holds significant potential both for facilitating the development of digital health innovations and for fostering their widespread adoption. In this study, we employed participatory action research (Baum et al., 2006) with PD as the primary research method. As summarized by Spinuzzi (2005), PD consists of three main stages: initial exploration of work, discovery processes, and prototyping.

As shown in Figure 8.1, we adapt PD for responsible innovation and apply it to the adoption phases within the digital health context. The adapted approach connects the three PD stages to the core considerations of responsible adoption. The first stage involves an initial exploration of the specific digital health innovations, including existing challenges, technology use, and potential integration into work practices. The second stage, referred to as the discovery process, focuses on gaining insights into stakeholders' perspectives, needs, and values in relation to digital health adoption. In the third stage, instead of following the conventional practice of using prototyping primarily for efficient and cost-effective design (Beaudouin-Lafon & Mackay, 2007), we emphasize stakeholder engagement to uncover the participatory process with responsible considerations.

Prior studies have identified various technical issues in digital health innovations, such as data acquisition accuracy, detection of abnormal results, and compatibility with patients' devices (Li et al., 2022, 2023). The focus of this study is specifically on revealing the human aspects

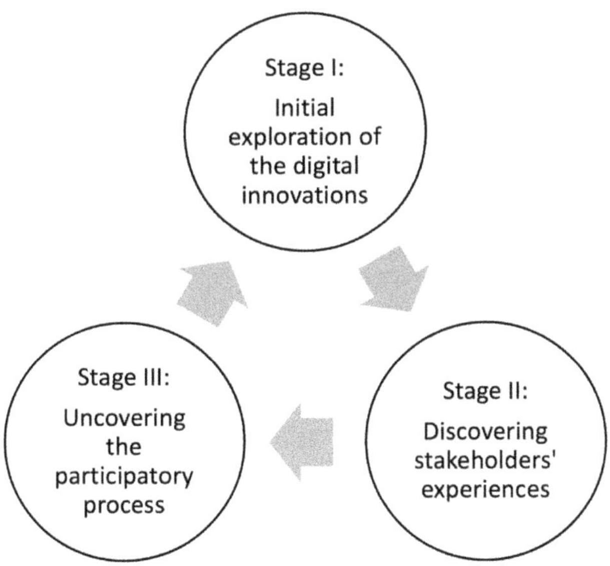

Figure 8.1 The responsible approach of participatory design (PD) in digital health innovation adoption.

associated with the adoption of digital health innovations. Across all stages, our approach remains anchored in responsible adoption practices, guided by the key dimensions of the RI framework: inclusion, anticipation, reflexivity, and responsiveness (Stilgoe et al., 2020).

Methodology

This study employed qualitative inquiry as its main research method (Glesne, 2016; Baskarada, 2014). Specifically, we investigated a case of digital remote care delivery at a tertiary hospital in China. Using a multi-method qualitative data collection approach, we examined the remote monitoring and rehabilitation services provided to 250 patients undergoing postoperative rehabilitation in a home setting. We sought to uncover the processes, challenges, and stakeholder experiences associated with implementing digital health innovations for the delivery of remote care services.

Based on the proposed PD approach (see Figure 8.1), the first author undertook onsite fieldwork during the rollout of digital health innovations for remote monitoring and rehabilitation. Table 8.1 outlines the data collection process, detailing the methods used, their purposes, durations, and locations. The data collection took place over three iterative PD stages, integrating the four key dimensions of the RI framework (Stilgoe et al., 2020). We operationalized the "inclusion" dimension during the data collection by engaging diverse stakeholders and empowering them to voice their opinions on shaping the responsible adoption of digital health innovations. Following data collection, we analyzed the data using qualitative content analysis (Elo & Kyngäs, 2008) to identify key themes and patterns that would facilitate the application of the proposed approach.

Table 8.1 Data collection across three iterative PD stages integrating the RI framework

PD stage	*Stage I*		*Stage II*	*Stage III*
RI dimension	Inclusion		Inclusion, reflexivity, anticipation	Inclusion, responsiveness
Key objective	Understanding the current status of the participatory process of the adoption.		Investigating direct stakeholders' experiences with the adoption	Unraveling the participatory process during the adoption
Data collection method	Observation	Document analysis	Semi-structured interview	Participatory workshop
Duration	One month	_	One hour with each health professional Half an hour with each patient	Two hours
Site	At the hospital	_	At the hospital with the health professionals Remotely with the patients via phone calls	At the hospital
Participants (with numbers)	Doctors, nurses, and patients (N/A)	_	Doctors ($N = 3$) Nurses ($N = 2$) Patients ($N = 10$)	Doctors ($N = 2$) Nurses ($N = 2$) Technology specialists ($N = 2$)
Documentation	Field notes and photos	Internal reports and documents	Audio recording	Audio recording

Applying PD to Shape the Responsible Adoption of Digital Health Innovations

Stage I: Initial Exploration of the Digital Health Innovations

We explored the participatory process for adopting digital health innovations through a relatively large-scale pilot study initiated by the thoracic surgery department of the hospital. The hospital implemented a suite of digital health innovations to deliver postoperative remote monitoring and rehabilitation services to 250 patients who had undergone thoracic surgery and were undergoing rehabilitation in a home setting.

As shown in Figure 8.2, the digital health innovations in our case included a web-based patient monitoring platform and a mobile application for patients connected to a range of vital sign measuring devices. The patient monitoring platform was developed for healthcare professionals to facilitate digital home follow-up, incorporating various patient-driven medical devices that support home-based care, including medical measurements, symptom reporting, and rehabilitation activities. The mobile application for patients was designed to be used in conjunction with connected devices, allowing patients to perform various vital sign measurements at home, such as blood pressure, pulse, temperature, oxygen saturation, lung function, etc. Measurement results were automatically recorded and transmitted via Bluetooth to the patient's end device, such as a smartphone. Simultaneously, these results were transferred to the patient monitoring platform, keeping health professionals updated with incoming values and enabling individualized follow-up care. The mobile application also enabled patients to monitor their vital signs, consult health professionals in real time, and respond to clinical questionnaires reporting symptoms and side effects.

Both health professionals and patients are direct stakeholders in the participatory process for adopting digital health innovations, embodying the "inclusion" dimension of RI. We recognize indirect stakeholders, such as technology vendors, patients' family members, caregivers, policymakers, and regulators, also contribute to the participatory process. Nevertheless, direct stakeholders are of particular interest in the next stage of the study due to their more immediate and substantial involvement in adopting digital health innovations.

Stage II: Discovering the Stakeholders' Experiences

By conducting semi-structured interviews with the direct stakeholders (i.e., three doctors, two nurses, and ten patients), we found their experiences in the pilot study were crucial for comprehending the intricacies of adopting digital health innovations. Our selection of these particular stakeholders was deliberate, owing to their direct interactions with the digital health innovations. Their experiences, captured in the interviews, supported a deeper analysis of the "inclusion", "reflexivity", and "anticipation" dimensions within the RI framework. Specifically, we aligned each of those dimensions with its unique characteristics:

- Inclusion: Emphasizing the different roles of the stakeholders and their respective adoption of digital health innovations.
- Reflexivity: Addressing their evolving perspectives and needs regarding digital health innovations.
- Anticipation: Highlighting the challenges and potential impacts they anticipated during the adoption of digital health innovations.

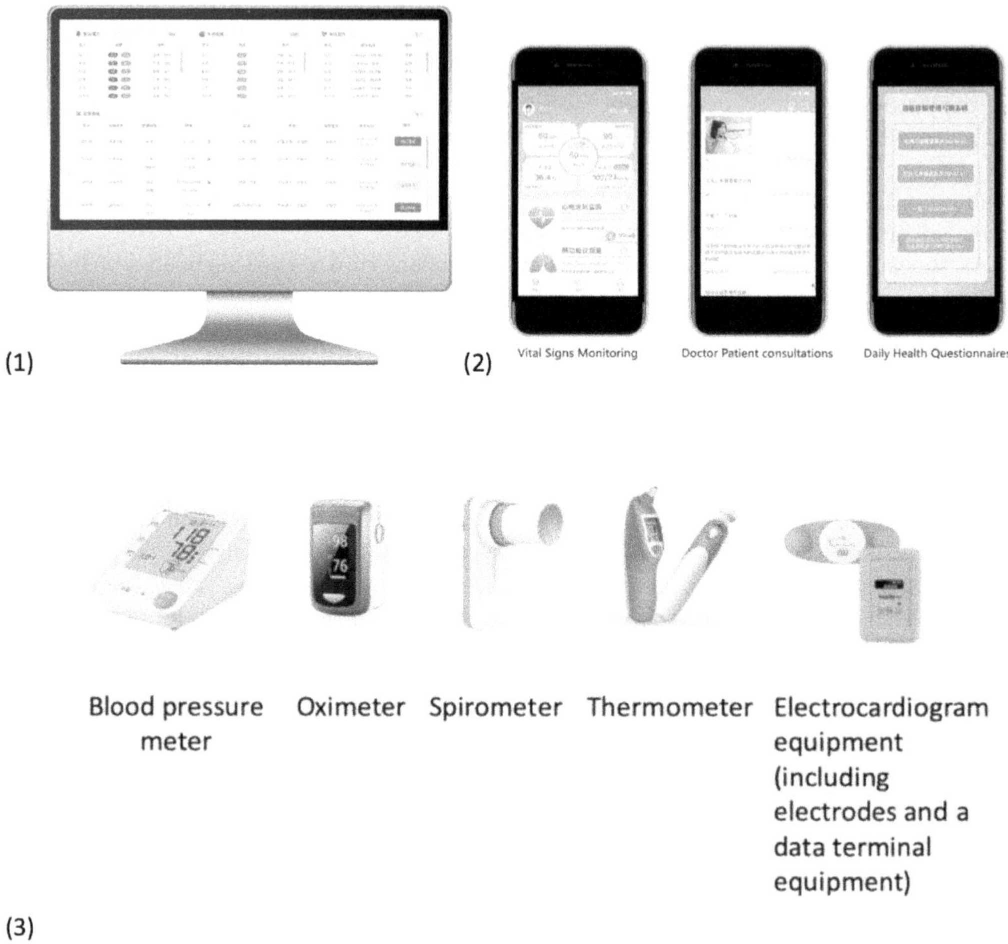

Figure 8.2 Overview of the digital health innovations used for postoperative remote care, including (1) a web-based patient monitoring portal for health professionals, (2) a mobile application for patients featuring vital signs monitoring and real-time consultations, and (3) connected vital signs measuring devices.

Inclusive Participation

To implement the digital health innovations, a dedicated team was formed, including surgeons, resident doctors, and nurses. This team played a pivotal role in patient inclusion in postoperative rehabilitation during the participatory process. They adopted the web-based patient monitoring portal to track the patients' rehabilitation progress, enabling real-time updates and communication. Assigned to digital follow-ups, nurses ensured continuous support and timely intervention when needed.

Our dataset revealed two distinct age groups among the 250 patients in the pilot study. The first group consisted of elderly patients aged 50–70, who often struggled with technology literacy and needed additional guidance and support to adopt the digital tools. In contrast, the second group, comprising younger patients aged 30–50, was generally more digitally fluent and found it easy to engage with the digital technologies for home-based postoperative rehabilitation. These age-related differences in digital literacy shaped how each group interacted with the digital health innovations, informing tailored approaches to patient engagement and care delivery within the study. These differences also guided our purposive sampling strategy, ensuring balanced representation by selecting patients for interviews from both age groups.

Reflective Insight

The health professionals in the team utilized digital health innovations to enhance patients' rehabilitation outcomes by continuously monitoring vital signs, tracking symptoms, and promoting adherence to prescribed rehabilitation plans. Compared to traditional follow-up services, which typically rely on phone calls to check on patients' status, health professionals noted several advantages of digital health innovations. They reflected that these tools could increase patient satisfaction and participation by providing more appropriate care and support beyond traditional remote care settings.

The ten patients interviewed identified several shared benefits from adopting the digital health innovations in their postoperative rehabilitation. They appreciated the ability to monitor vital signs using connected devices, receive care at home, and communicate with healthcare professionals via the mobile application. This reduced the need for frequent hospital visits. It was particularly valuable for those living in remote areas and during the COVID-19 pandemic when mobility was restricted. Most patients also felt reassured knowing their postoperative status was continuously monitored remotely. It gave them confidence that any potential complications could be detected early, reducing the risk of delayed treatment. In addition, many reported feeling empowered by actively participating in their own care. This was particularly beneficial for elderly patients, as it offered tools to better manage chronic conditions. For younger patients, the ability to track their rehabilitation progress instilled a sense of control and motivated them to preserve their independence and autonomy throughout the postoperative rehabilitation process.

Anticipatory Impact

Despite the benefits of digital health innovations, the healthcare professionals pointed out that these tools cannot fully replace the emotional and psychological support patients typically need in face-to-face care during rehabilitation. Although the team worked to integrate the digital health innovations into their work practices to enhance the patient experience, they encountered several challenges, including heavy workloads and a lack of interdepartmental collaboration. Relying solely on the dedicated remote care team was viewed as insufficient to deliver optimal outcomes in digital remote care. The health professionals also expressed concerns about the absence of guidelines, policies, and regulations to promote the responsible adoption of digital health innovations.

Interviews with the selected patients revealed that both groups faced distinct challenges that could potentially affect compliance, which was a primary concern in the pilot study. For elderly patients, limited digital literacy and physical impairments (e.g., arthritis, age-related vision decline, postoperative discomfort) hindered their ability to adopt digital health innovations. Despite receiving training before discharge, patients with low technological exposure or cognitive impairments (e.g., dementia) had difficulty engaging and often required repetitive training. The absence of social support from family or caregivers further hindered some patients' ability to adopt digital health innovations. Emotional barriers also emerged, particularly among elderly patients accustomed to traditional in-person care. A concern raised by patients was that digital health innovations could compromise the quality of interactions with their healthcare providers. They feared it might replace the personal touch and relationship they had established. In contrast, younger patients encountered fewer technical difficulties but showed limited awareness of the importance of postoperative rehabilitation. This issue was particularly evident among first-time thoracic surgery patients, who often found postoperative rehabilitation emotionally challenging. The digital health innovations adopted in the pilot study lacked built-in mechanisms to incentivize or reinforce the rehabilitation progress, which might lead to decreased patient engagement.

Stage III: Uncovering the Participatory Process

The prior analysis of the "inclusion", "reflexivity", and "anticipation" dimensions based on direct stakeholders' experiences was essential for informing future design initiatives that address the raised concerns and align with the stakeholder needs. Building on this foundation, we proceeded to gain a holistic view of the participatory process of adopting digital health innovations from direct stakeholders' perspectives. This stage was guided by the "inclusion" and "responsiveness" dimensions of the RI framework. We held a participatory workshop with doctors, nurses, and technology specialists, using collective reflexivity to shape the innovation pathway and set the direction for their responsible adoption.

The workshop participants co-designed the participatory process for adopting digital remote care in postoperative rehabilitation. Figure 8.3 illustrates a sequential progression across key phases, from pre-surgery preparation to after-discharge monitoring, while ensuring responsible implementation. The workshop facilitated discussions on adoption challenges, improvement areas, and stakeholder alignment around a shared vision in the participatory process.

Due to ethical restrictions, direct patient participation in the workshop was prevented. However, we found the nurses played a dual role in the workshop, offering their professional insights while representing patients' voices. Their empathy for patients and firsthand knowledge helped incorporate patient perspectives into discussions and decision-making during the workshop. In this case, the doctors acted as both users and managers of the patient monitoring platform. They provided valuable insights into the platform's usability and effectiveness during the workshop. The technology specialists offered suggestions and potential remedies for addressing technology-related issues that surfaced during the implementation. Furthermore, by developing an empathic understanding of patients and health professionals through insights shared during the workshop, the technology specialists identified valuable input to guide future iterations of the digital health innovations.

Discussion

Drawing upon the findings of our case study, we identified potential pitfalls that may impede the responsible adoption of digital health innovations. Furthermore, we propose recommendations for future design initiatives to foster a more responsible implementation.

Pitfalls in Promoting a Responsible Adoption of Digital Health Innovations

Based on the data from stakeholder interviews and the workshop discussions, we collectively identified a series of interconnected challenges during the participatory process of responsibly adopting digital health innovations. We further categorized these challenges into four levels of

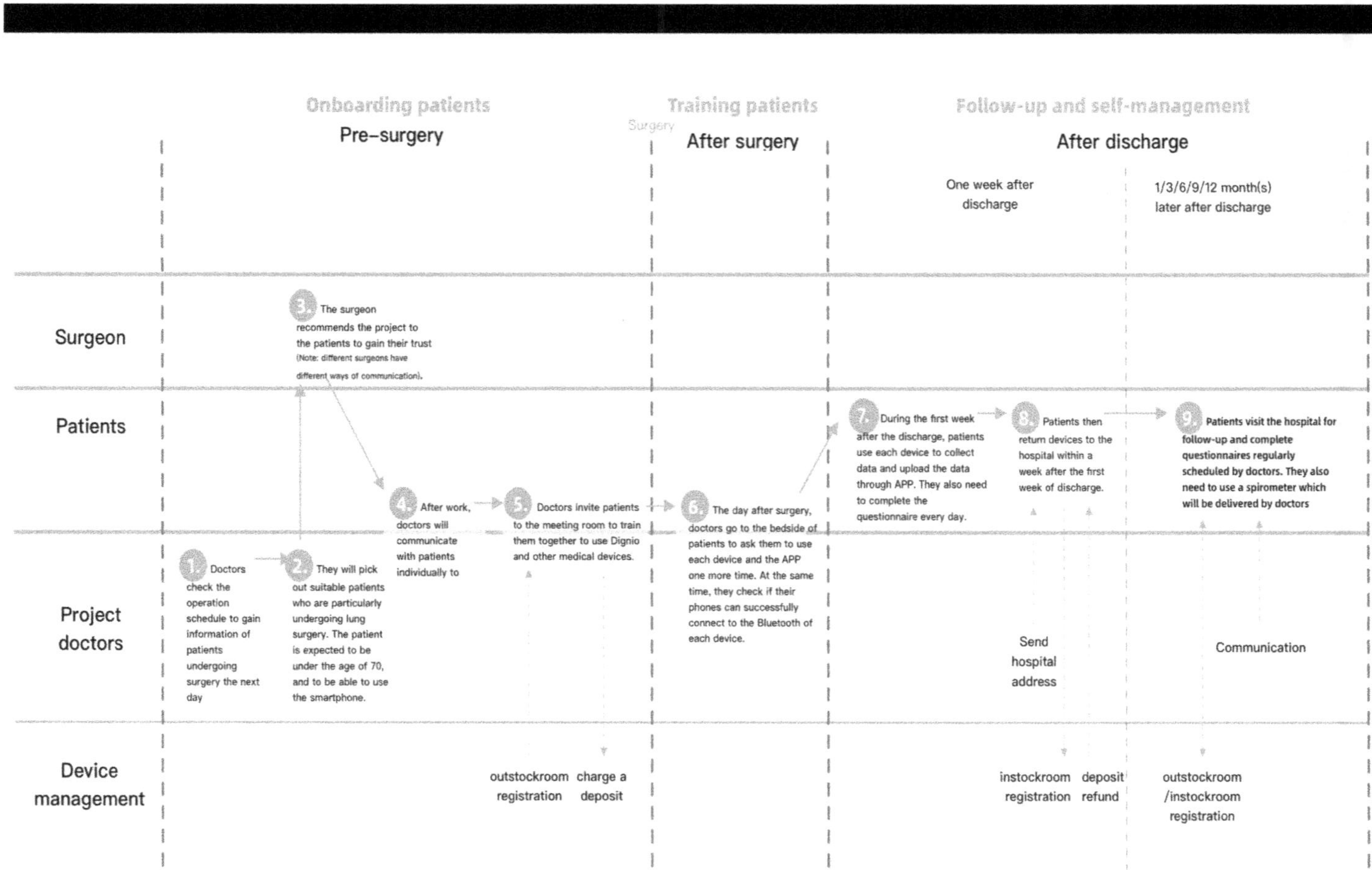

Figure 8.3 A reproduced illustration of the participatory process for adopting digital remote care in the case study.

Figure 8.4 The four-level pitfalls in responsible adoption of digital health innovations.

pitfalls, affecting stakeholders from the individual level to broader systemic levels, as illustrated in Figure 8.4.

At the first level, patients as direct users of digital health innovations faced barriers linked to limited understanding of their condition and the role of postoperative rehabilitation. Our case study revealed that such knowledge gaps often lead to a lack of awareness about the benefits of digital health innovations. Without adequate training in essential skills, patients might struggle to effectively adopt digital health innovations to track their rehabilitation progress and comprehend the impact of their condition.

This leads to the second level of pitfalls regarding patient motivation to engage with digital health innovations. Engagement appeared to depend heavily on having the necessary knowledge, awareness, and skills to use them meaningfully. Notably, the adoption of digital health innovations emerged as a learning process for both patients and health professionals in the pilot study. Patients needed to learn how to use the technologies to support their postoperative rehabilitation at home, while health professionals had to adapt their clinical routines and integrate the tools into their existing workflows. This dual learning process highlighted key integration challenges, particularly the need to align digital technologies with the lived realities and expectations of different stakeholders. Adoption was further complicated by a lack of seamless integration with existing health information systems, workflow disruptions, and role ambiguity. These issues underscore the need for a more ecosystem-aware design approach—one that accounts not only for the technical usability of a digital tool but also for its socio-organizational embedding.

Organizational-level collaboration is crucial for resolving the third-level pitfalls identified in this study. In the pilot study, only a small remote care team within a single hospital department implemented the digital health innovations, which limited the available human resources for supporting remote monitoring and postoperative rehabilitation for a group of 250 postoperative patients. For more responsible adoption, fostering internal collaboration and leveraging the expertise of related hospital departments will be crucial.

On the governmental level (i.e., the fourth level), we identified a lack of standard guidelines for the responsible adoption of digital health innovations, along with policy and regulatory gaps pertaining to postoperative care responsibilities. Across all identified levels, addressing the pitfalls requires coordinated action and broader stakeholder engagement. Such collaborative efforts essentially involve direct stakeholders at the individual level, such as doctors, nurses, and patients, who engage firsthand with digital health innovations and play a crucial role in shaping their practical implementation and clinical effectiveness. Meanwhile, indirect stakeholders, such as hospital administrators, healthcare policymakers, and regulatory authorities, contribute to the broader systemic framework by establishing guidelines, allocating resources, and ensuring compliance with ethical and legal standards. Synergies between direct and indirect stakeholders are essential for bridging the gap between frontline needs and overarching institutional decision-making. Ultimately, strong organizational and governmental support, coupled with strategic decision-making, is key for guiding future design initiatives that facilitate the responsible and sustainable adoption of digital health innovations.

Recommendations for More Responsible Adoption of Digital Health Innovations

We propose a set of recommendations for responsible adoption of digital health innovations, drawing on the key principles of RI (von Schomberg, 2013), specifically those closely linked with organizational decision-making, i.e., *(ethical) acceptability, sustainability,* and *societal desirability*. We operationalize these dimensions in the specific context of adopting digital health innovations. Informed by technology acceptance models, our recommendations place particular emphasis on four factors relevant to healthcare settings: "perceived usefulness", "perceived ease of use", "social influence/subjective norms", and "perceived behavioral control/facilitating conditions" (Davis, 1989; Venkatesh & Davis, 2000; Venkatesh et al., 2003; Holden & Karsh, 2010).

First, "perceived usefulness" captures the idea that digital health technologies must be designed to deliver meaningful and functional benefits to their direct users. For instance, in the context of digital remote, providing real-time feedback on patients' rehabilitation progress enables health professionals to make informed decisions and deliver personalized care promptly. Furthermore, these technologies should be designed with patient-centric value in mind, enabling patients to understand their medical data, track their progress, and develop the skills necessary to effectively manage their conditions. Empowering patients helps transform them from passive recipients of healthcare services into active participants by fostering self-management skills (Anderson & Funnell, 2010). However, it is essential to acknowledge the complexities and ambivalence involved in empowering a layman to adopt digital technologies for health and illness management. As Lupton (2013) noted, encouraging patients to adopt digital health innovations requires addressing the surveillance and disciplinary aspects of self-monitoring tools, the emotional responses and resistance they may provoke, the increased burden of self-care, and the often-overlooked labor undertaken by healthcare professionals to support their use.

Second, "perceived ease of use" refers to the degree of ease associated with adopting digital health innovations, with the aim of minimizing the physical or mental effort required to use them (Davis, 1989; Venkatesh et al., 2003). In a digital remote care environment, both health professionals and patients are users of the digital health technologies. The solution should provide clear and easy-to-understand instructions on how to use it, along with ongoing training and support to enhance users' confidence and competence in adopting the technologies. A user-friendly design with a low learning curve is critical for individuals with limited technological literacy to engage meaningfully with the technology. For health professionals, the technologies should integrate seamlessly into existing workflows to support adaptability and minimize disruptions to their practice. For patients, embedding these technologies within familiar and mainstream systems already part of their daily routines can facilitate their ease of use.

Third, "social influence/subjective norms" reflect how the perceptions and opinions of broader stakeholder groups can influence the adoption of digital health innovations (Holden &

Karsh, 2010). For instance, patients are more likely to adopt new technologies when endorsed by healthcare professionals, as such endorsements raise awareness and perceived benefits. Additionally, involving family, friends, and peers in the adoption process can strengthen social support, thereby increasing patient motivation and engagement. We argue that embedding societal desirability into the adoption of digital health innovations requires collaboration across multiple levels (WHO, 2021). In particular, addressing the identified organizational and governmental pitfalls calls for cross-departmental integration within hospital systems. This is essential for seamless integration of these innovations into existing workflows and care processes. Equally vital is governmental support. It plays a key role in establishing guidelines, standards, policies, and regulations for responsible adoption. They safeguard patient data privacy and security, guide ethical practices, and facilitate the responsible large-scale implementation of digital health innovations.

Fourth, "perceived behavioral control/facilitating conditions" entail the perception that the necessary organizational and technical infrastructure is in place to support technology adoption (Venkatesh et al., 2003). While user acceptance is critical, a smooth transition to digital remote care also requires a robust technical infrastructure compatible with existing healthcare systems. However, the four levels of pitfalls identified in this study reveal that technical readiness alone is not enough. This highlights the need for an ecosystem-aware design approach sensitive to complex social and organizational contexts. Sustaining the implementation of digital health innovations demands mutual adaptation. The technology must align with users' everyday environments (both patient and clinical), and those environments must, in turn, be reshaped to accommodate new digital practices. This interplay constitutes a dynamic and reciprocal process. A robust technical foundation is essential, but it must be coupled with participatory processes and adaptive strategies that account for the realities of clinical practice and the lived experiences of patients.

Reimagining Engagement through the Integrated Lens of PD and RI

Our study reimagined engagement by moving beyond the early stages of technological innovation development, where user engagement is typically concentrated, to focus on the post-innovation phase. This phase, where the evaluation and adoption process is critical, is often overlooked in PD or RI discourses. By embedding the RI framework within a PD approach, we offer a practical way to engage stakeholders in the use of digital health innovations. This temporal shift emphasizes ongoing engagement and iterative design for its continuous improvement, rather than one-off consultations or user tests that may not fully capture real-world use (Bødker et al., 2017; Owen et al., 2020).

The originality of this approach lies in its capacity to surface persistent and systemic issues that may not be visible during initial development stages. While some of the pitfalls identified might seem self-evident in hindsight, uncovering them through a structured framework that integrates PD and RI adds both methodological rigor and analytical credibility to the findings. More importantly, it points to the value of structured reflection and stakeholder (re-)engagement even after technologies are deployed, an area largely neglected in current digital health literature (Greenhalgh et al., 2017).

Through the integrated lens of PD and RI, we reconceptualize engagement not as a linear or closed process, but as an ongoing dialogue. It promotes a dynamic mode where engagement is distributed across the lifecycle of technological innovations. Such a mode allows for feedback loops and adaptation of digital solutions based on lived experiences and contextual realities. This chapter champions designing with stakeholders to support more meaningful and responsive innovation (Stilgoe et al., 2020). Our real-world validation of the RI framework within PD in the digital remote care setting provides a basis for future research and tailored implementations elsewhere. As it is extended into new contexts and populations, it offers opportunities for richer insights and more grounded implications.

Conclusion

In this chapter, we have deepened our understanding of a responsible PD approach to reimagining engagement, one that involves stakeholders more effectively, fosters empathy, and empowers them to contribute to the responsible adoption of digital health innovations. This study provides three main contributions. First, it introduces a novel connection between PD and RI at the post-innovation stage, particularly during the adoption of digital health innovations. In doing so, it proposes a responsible PD approach that broadens PD's traditional scope of application.

Second, we provided empirical evidence from a real-world case, demonstrating that applying a responsible approach to PD can support the adoption of digital health innovations. Our findings show this approach as an appropriate working method for bottom-up exploration of multistakeholder engagement toward more responsible adoption. In the case study, we reached common ground on "inclusion" between the RI framework and PD by engaging social actors (i.e., the direct stakeholders in our case) and incorporating their diverse perspectives into the three PD stages. The "anticipation" and "reflexivity" dimensions of the RI framework guided the investigation in the discovery stage, while the "responsiveness" dimension informed the direction for addressing the identified challenges.

Third, we identified key pitfalls in responsible digital health adoption and offered corresponding recommendations based on RI principles and technology acceptance models. These recommendations are intended to guide future design initiatives toward more responsible digital health adoption. To broaden the applicability of our findings beyond postoperative remote care, we stress the critical need for an "all-inclusive multistakeholder approach" (WHO, 2021). Future adoption efforts should be both responsible and participatory, tailored to varied contexts.

In closing, our work promotes a new mode of engagement: one that is reflective, recursive, and grounded in the complexities of actual implementation. This, we argue, is where true co-creation and participatory impact can be realized, not only in design but also in the long-term shaping and reshaping of digital health innovations.

Acknowledgments

This study is part of the DigiRemote project (No. 310137), which received funding from IKTPLUSS-IKT and a digital innovation grant from the Research Council of Norway.

References

Anderson, Robert M., and Martha M. Funnell. "Patient Empowerment: Myths and Misconceptions." *Patient Education and Counseling* 79, no. 3 (2010): 277–282.

Barlow, Julie, Chris Wright, Janice Sheasby, Andy Turner, and Jenny Hainsworth. "Self-Management Approaches for People with Chronic Conditions: A Review." *Patient Education and Counseling* 48, no. 2 (2002): 177–187.

Baskarada, Sasa. "Qualitative Case Study Guidelines." *The Qualitative Report* 19, no. 40 (2014): 1–25.

Baum, Fran, Colin MacDougall, and Danielle Smith. "Participatory Action Research." *Journal of Epidemiology and Community Health* 60, no. 10 (2006): 854.

Beaudouin-Lafon, Michel, and Wendy E. Mackay. "Prototyping Tools and Techniques." In *The Human-Computer Interaction Handbook*, edited by Andrew Sears and Julie A. Jacko, 2nd ed., 1043–1066. CRC Press, 2007.

Bentham, Ashleigh. "Design for Next Thinking: Using Transformation Consequence Mapping to Improve Responsible Innovation Practices." *The Design Journal* 20, no. sup1 (2017): S4281–S4291.

Bødker, Susanne, Christian Dindler, and Ole Sejer Iversen. "Tying Knots: Participatory Infrastructuring at Work." *Computer Supported Cooperative Work (CSCW)* 26 (2017): 245–273.

Craigon, Peter J., Justin Sacks, Steve Brewer, Jeremy Frey, Anabel Gutierrez, Naomi Jacobs, Samantha Kanza, et al. "Ethics by Design: Responsible Research & Innovation for AI in the Food Sector." *Journal of Responsible Technology* 13 (2023): 100051.

Davis, Fred D. "Perceived Usefulness, Perceived Ease of Use, and User Acceptance of Information Technology." *MIS Quarterly* 13, no. 3 (1989): 319–340.

Elo, Satu, and Helvi Kyngäs. "The Qualitative Content Analysis Process." *Journal of Advanced Nursing* 62, no. 1 (2008): 107–115.

Glesne, Corrine. *Becoming Qualitative Researchers: An Introduction.* Pearson, 2016.

Greenhalgh, Trisha, Joseph Wherton, Chrysanthi Papoutsi, Jennifer Lynch, Gemma Hughes, Susan Hinder, Nick Fahy, Rob Procter, and Sara Shaw. "Beyond Adoption: A New Framework for Theorizing and Evaluating Nonadoption, Abandonment, and Challenges to the Scale-Up, Spread, and Sustainability of Health and Care Technologies." *Journal of Medical Internet Research* 19, no. 11 (2017): e8775.

Holden, Richard J., and Ben-Tzion Karsh. "The Technology Acceptance Model: Its Past and Its Future in Health Care." *Journal of Biomedical Informatics* 43, no. 1 (2010): 159–172.

Kipnis, Jeffrey. "Towards a New Architecture: Folding." *Architectural Design* 63 (1993): 3–4.

Lehoux, Pascale, Hudson Pacifico Silva, Renata Pozelli Sabio, and Federico Roncarolo. "The Unexplored Contribution of Responsible Innovation in Health to Sustainable Development Goals." *Sustainability* 10, no. 11 (2018): 4015.

Li, Hong, Torbjørg Meum, and Miria Grisot. "Exploring Early-Stage Implementation of Digitally Enabled Remote Care: Case Studies from Norway and China." In *Scandinavian Conference on Health Informatics*, 2–8. Linköping University Electronic Press, 2022.

Li, Hong, Torbjørg Træland Meum, and Miria Grisot. "Introducing Digital Technologies for Remote Care in Norway and China: The DigiRemote Project." *Procedia Computer Science* 219 (2023): 1478–1484.

Lupton, Deborah. "The Digitally Engaged Patient: Self-Monitoring and Self-Care in the Digital Health Era." *Social Theory & Health* 11 (2013): 256–270.

Lynn, Greg. "Architectural Curvilinearity, the Folded, the Pliant and the Supple." *Architectural Design* 102 (1993): 8–15.

Owen, Richard, Phil Macnaghten, and Jack Stilgoe. "Responsible Research and Innovation: From Science in Society to Science for Society, with Society." In *Emerging Technologies*, edited by Gary E. Marchant and Wendell Wallach, 117–126. Routledge, 2020.

Oxman, Rivka. "Theory and Design in the First Digital Age." *Design Studies* 27, no. 3 (2006): 229–265.

Pavie, Xavier, and Daphné Carthy. "Leveraging Uncertainty: A Practical Approach to the Integration of Responsible Innovation through Design Thinking." *Procedia – Social and Behavioral Sciences* 213 (2015): 1040–1049.

Pavie, Xavier, Daphné Carthy, and Victor Scholten. *Responsible Innovation: From Concept to Practice.* World Scientific, 2014.

Silva, Luciana Maines da, Claudia Cristina Bitencourt, Kadígia Faccin, and Tatiana Iakovleva. "The Role of Stakeholders in the Context of Responsible Innovation: A Meta-Synthesis." *Sustainability* 11, no. 6 (2019): 1766.

Spinuzzi, Clay. "The Methodology of Participatory Design." *Technical Communication* 52, no. 2 (2005): 163–174.

Stilgoe, Jack, Richard Owen, and Phil Macnaghten. "Developing a Framework for Responsible Innovation." In *The Ethics of Nanotechnology, Geoengineering, and Clean Energy*, edited by Andrew Maynard and Jack Stilgoe, 347–359. Routledge, 2020.

van Berkel, B., Caroline Bos, and UNStudio. *Move.* 1st ed. UN Studio & Goose Press, 1999.

Venkatesh, Viswanath, and Fred D. Davis. "A Theoretical Extension of the Technology Acceptance Model: Four Longitudinal Field Studies." *Management Science* 46, no. 2 (2000): 186–204.

Venkatesh, Viswanath, Michael G. Morris, Gordon B. Davis, and Fred D. Davis. "User Acceptance of Information Technology: Toward a Unified View." *MIS Quarterly* 27, no. 3 (2003): 425–478.

von Schomberg, Rene. "A Vision of Responsible Research and Innovation." In *Responsible Innovation: Managing the Responsible Emergence of Science and Innovation in Society*, edited by Richard Owen, John Bessant, and Maggy Heintz, 51–74, 2013. John Wiley & Sons.

World Health Organization. *Global Strategy on Digital Health 2020–2025*. World Health Organization, 2021. Accessed June 28, 2025. www.who.int/docs/default-source/documents/gs4dhdaa2a9f352b0445bafbc79ca799dce4d.pdf.

9 Usability in Sustainable Systems

A Service Design Perspective

Kuisma Hurtig, Timo Luukkonen, Kati Pitkänen, and Ashley Colley

Introduction

Citizen participation in sustainability efforts depends not only on awareness or willingness to act but also on how easily sustainable actions can be integrated into everyday life. When it's about sustainable commitment and more about system-supported behavior. This chapter explores how design can lower barriers to participation by embedding reuse into ordinary consumer interactions.

Finland's beverage bottle and can deposit-return system is one of the most comprehensive in the world. As of 2023, over 5,000 reverse vending machines were distributed across grocery stores, kiosks, and service stations (Finland Promotion Board, 2023). In 2024, a record 2.3 billion containers were returned, reflecting the system's widespread acceptance (Helsinki Times, 2024). Deposit values vary by size and type, from €0.10 for small plastic bottles to €0.40 for large ones (Institute for European Environmental Policy (IEEP), 2022). The system is deeply embedded in Finnish everyday life and is considered a model for citizen-engaged recycling.

In contrast, single-use plastic shopping bags remain a persistent environmental concern, contributing to waste, pollution, and ecosystem harm (United Nations Environment Programme, 2018). Although policy interventions, such as bag fees, have led to significant reductions in use (e.g., a 97% drop in the UK after a 5p charge (Behavioural Insights Team, 2022), consumption remains high in many countries. The EU has set a target of 40 bags per person annually by 2025 (European Union, 2015), and Finland has made progress via voluntary agreements and pricing, reducing use to around 56 bags per person (Ympäristöministeriö, 2023). However, stronger measures are needed to shift from single-use to reusable options.

Although recycling receives policy emphasis, reuse sits higher in the waste hierarchy and offers greater environmental benefits by reducing both material input and energy use (European Environment Agency, 2020). However, reuse systems often face adoption challenges. When perceived as inconvenient, unfamiliar, or disconnected from existing routines, they are easily ignored. Shopping bags provide a valuable design case: they are part of daily shopping, already regulated, and embedded with both symbolic and practical functions. Common reasons for choosing single-use bags include forgetting to bring a reusable one, underestimating need, or planning to use them as trash bags (Pitkänen et al., 2024).

While behavioral and regulatory perspectives on reuse are well explored, fewer studies examine how everyday familiar infrastructures might be repurposed to support reuse practices in public consumer contexts. This chapter addresses that gap by asking: How can routine-based, familiar interaction models be used to support citizen engagement with reusable packaging systems? We approach reuse not only as a behavioral or policy problem but as a service design challenge that involves shaping systems to feel intuitive and socially embedded.

DOI: 10.4324/9781003667872-12

Our approach is grounded in design research that treats services as infrastructural arrangements that mediate participation (Manzini, 2015) and that sustainable behavior change must be embedded in user experiences that are both intuitive and meaningful (Kuijer & Bakker, 2015; Meroni & Sangiorgi, 2011). In this chapter, we understand engagement not as a fixed internal trait such as motivation or interest, but as something that emerges from the alignment of systems, routines, and user expectations. Engagement is shaped through design decisions that reduce friction, connect to familiar practices, and support ease of participation (Kuijer & Bakker, 2015; Wever et al., 2008).

To explore these ideas, we conducted an online survey to gather general attitudes toward reusable products with a deposit model (n = 84), followed by the development and evaluation of a prototype service for reusable shopping bags based on Finland's well-known bottle return system. The prototype was tested through guided user walk throughs (n = 24), allowing participants to experience and reflect on the service path. Participants responded positively to the model's familiarity and ease of use, though usability issues were identified that could reduce adoption.

This chapter contributes to discussions on citizen-centered sustainability by showing how design methods can identify new ways to embed reuse into everyday life.

Related Work

This study builds on interdisciplinary research at the intersection of sustainability transitions, service design, participatory approaches, and design research methods. To position our work, we review prior work focusing on how everyday infrastructures and routines have been mobilized to support reusable systems and where gaps remain in understanding engagement as a designable quality of interaction.

Sustainability, Plastic Waste, and Reuse Systems

Research on sustainability transitions has emphasized the importance of shifting everyday practices through systems that extend product lifetimes. While plastic bag charges have proven effective in reducing single-use consumption (Jakovcevic et al., 2014), long-term change depends on meaningful alternatives that people can engage with in daily life. Reuse, in particular, requires repeated participation and user-friendly infrastructures that support continuity of use (Wever et al., 2008). For example, Closed Loop Partners (2025) reports on a multi-retailer bring-your-own-bag pilot in the United States, showing how reuse can scale when embedded in existing retail routines.

Recent policy frameworks are increasingly mandating such reuse systems. The European Union's Regulation (EU) 2025/40 outlines stricter requirements for reusable packaging across sectors, not just for beverages or transport packaging, signaling a shift toward infrastructure-based circularity (European Union, 2024).

However, much of this research has focused on high-level policy outcomes, technical feasibility, or consumer intentions. Less attention has been given to how reuse can be supported through familiar infrastructures that minimize friction and normalize participation. We focus on these everyday service interactions and ask how reuse can be designed to feel as intuitive and habitual as current return systems.

Service Design and Everyday Participation

Within design research, service design has been positioned as a means of enabling sustainable practices by aligning them with familiar routines. Meroni and Sangiorgi (2011) argue that service design can support social innovation through accessible, human-scale interventions. Kuijer and Bakker (2015) further emphasize that successful interventions must connect to existing material practices to ensure adoption.

Several recent studies have explored how users perceive and interact with reusable packaging systems. Miao et al. (2023) found that ease of use, hygiene, and system clarity strongly influence participation. Tenhunen-Lunkka et al. (2024) argue that interdisciplinary service design is needed to operationalize circular packaging, based on a reusable pizza box case. Espinoza-Orias and Lundquist (2025) identify best practices that combine technical reuse metrics with user experience considerations in food and beverage contexts.

While these works support the importance of design in sustainable systems, few have investigated how specific interaction models, such as deposit-return systems, can be adapted to new domains like shopping bags. Our work builds on this by exploring how familiar return mechanisms can be extended to support reuse in a different product category and how this influences perceived usability and participation.

Participatory Design and Infrastructure Building for Sustainability

While user-centered and service design approaches shape how reuse systems are implemented, participatory design (PD) addresses how citizens contribute to their formation and ongoing development. Originally grounded in workplace democratization (Björgvinsson et al., 2012), PD has evolved to tackle broader systemic challenges, including sustainability and public sector innovation. More recent work highlights its role in shaping long-term infrastructures through iterative engagement (Le Dantec & DiSalvo, 2016).

We approach PD here not through co-design workshops, but via low-fidelity prototypes and service walkthroughs used as reflective tools. These allow participants to engage directly with a proposed future system—critiquing, interpreting, and imagining it in use.

Although participatory approaches are often applied at early planning stages, fewer studies examine how interaction with speculative service prototypes can generate actionable feedback. Our study extends this space by using physical artifacts to elicit situated responses and highlight frictions in reuse service design.

Design Research Methods for Exploring Use and Adoption

Methodologically, this study draws from design research traditions that treat prototypes not only as validation tools but also as instruments of inquiry. Research-through-design (RtD) employs speculative artifacts to generate insights through reflective user engagement (Zimmerman et al., 2007). In HCI and service design, methods such as walkthroughs, journey maps, and touchpoint simulations are used to surface user expectations, highlight breakdowns, and support iteration (Colley et al., 2021; Stickdorn et al., 2018).

Our study follows this tradition by applying staged, low-fidelity interaction in a lab setting to examine a routine-based reuse system. While most RtD studies focus on novel technologies, we apply this approach to the mundane context of reusable shopping bags to show how users make sense of reuse services embedded in their everyday routines. This helps us understand how trust, convenience, and acceptance are shaped not just by features, but by design cues grounded in familiarity.

Method: Survey and Experience Prototyping

To explore how routine-based service interactions support engagement with reusable packaging, we conducted a two-part study. An online survey gathered attitudes toward deposit-return systems, followed by the design and evaluation of a service path and low-fidelity prototype for a reusable shopping bag return system.

Online Survey

An online survey was developed to explore public opinions on deposit-return packaging systems. The survey began with background questions about respondents' current experiences with existing deposit-return schemes, such as those for bottles and cans. Respondents were then introduced to three proposed concepts: a reusable shopping bag, a takeaway food container, and a fresh produce bag. They were asked to assess each concept's feasibility and suggest an appropriate deposit amount. The survey also invited participants to identify other product types they believed could be suitable for deposit-return and to reflect on the main challenges associated with such systems. Finally, respondents could contribute additional ideas or comments related to the topic.

Participants. Survey participants were recruited from a citizen panel of 138 individuals assembled by the Finnish Environment Institute (Syke), comprising people from across the country who had expressed a general interest in plastic recycling and concern about plastic pollution. Most of the panelists could be described as environmentally conscious citizens. The survey was distributed to the panel via email and received 84 responses. Of the respondents, 71 identified as female, 11 as male, and 2 as other. Age distribution was as follows: 4 participants were under 26 years old, 19 were aged 26–35, 22 were aged 36–45, 25 were aged 46–55, 12 were aged 56–65, and 3 were aged 66 or older.

Service Concept and Evaluation

Based on the survey results, we developed a service concept for a reusable shopping bag deposit-return system (Figure 9.1). A low-fidelity prototype of the return machine was created, allowing study participants to experience a key touchpoint of the service in a controlled setting (Figure 9.2). The paper-based interface was manually operated by a facilitator during participant interaction.

While the study followed a user-centered structure, the evaluation incorporated PD principles. In addition to completing predefined tasks, participants were invited to interpret and critique the system, aligning with PD approaches that treat users as active interpreters of future possibilities (Björgvinsson et al., 2012; Le Dantec & DiSalvo, 2016).

Participants first reviewed a visualized service path (Figure 9.1), which outlined all key steps of the system, including obtaining, using, and returning the bags. Afterwards, they completed a short questionnaire to record initial impressions. In the second part of the evaluation, participants used the return machine prototype to complete a realistic task: returning three shopping bags. They were asked to think aloud during the interaction (Ericsson & Simon, 1993). Observers recorded both verbal feedback and observable behavior. After the task, participants completed a questionnaire with semantic differential scales (e.g., 'practical–impractical') and open-ended items focused on usability and challenges. This setup allowed us to explore how familiarity and routine influence engagement in a speculative reuse system, addressing the study's main research question.

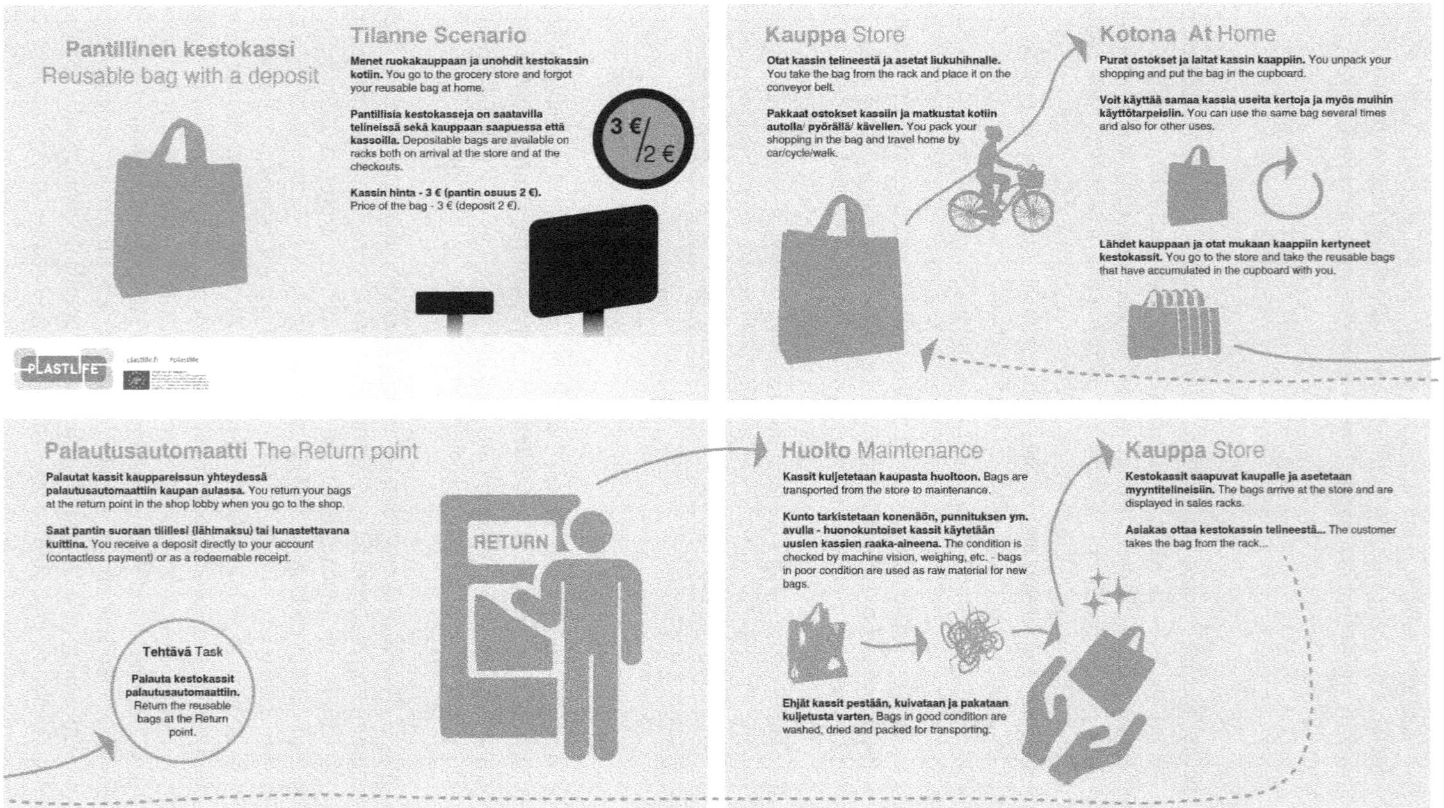

Figure 9.1 Visualized service path of the reusable bag concept (Finnish and English text).

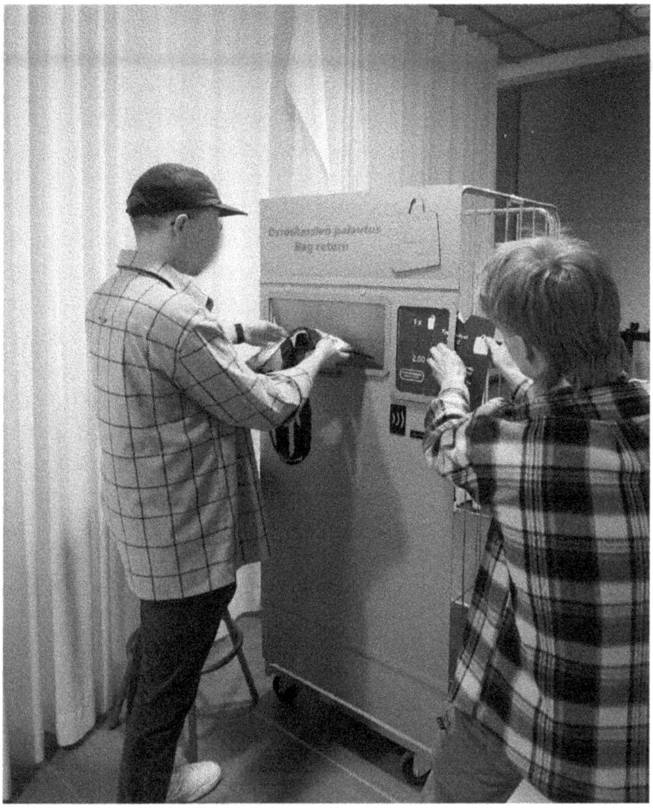

Figure 9.2 Participant completing the returning task with a cardboard prototype.

Participants. A total of 24 service design students participated in the experience prototyping study. The gender distribution was as follows: 15 identified as women, 7 as men, 1 as other, and 1 preferred not to disclose. Participants' age distribution was as follows: 9 were aged 18–25, 12 were aged 26–35, 1 was aged 36–45, and 1 was aged 46–55. Eighteen participants reported being highly familiar with the Finnish bottle return system, which served as the model for the proposed service, while the remaining 6 participants had varying levels of familiarity.

Results

This section presents the findings from the two stages of the study. The first subsection reports survey responses related to public perceptions of deposit-return systems and reusable packaging. The second presents findings from the service path walkthrough and return machine prototype evaluation, focusing on how participants experienced and interpreted the proposed concept.

Online Survey

The survey results offer insights into citizen attitudes toward deposit-return systems, perceived barriers to reuse, and the relative appeal of various packaging types. These findings informed the development of the service path and prototype.

The national bottle and can return system is well known in Finland and was familiar to all survey respondents (Helsinki Times, 2024). However, when asked whether they had tried deposit-based packaging for products other than bottles, 86% reported either not doing so or not being aware that such systems existed. Only 14% had first-hand experience, citing examples such as deposit cups at music festivals or returnable packaging from online retailers.

Respondents expressed generally positive attitudes toward deposit-based systems. When asked about expanding such systems beyond beverage packaging, the average support rating was 4.2 out of 5. Still, 65 out of 84 respondents identified consumer attitudes as a major barrier to broader adoption. Other key enablers mentioned were regulation and financial incentives, likely reflecting the environmental awareness of the panel, which may not generalize to the broader population.

Logistics, consumer attitudes, and hygiene were the most frequently selected challenges (Figure 9.3). Additional concerns included packaging design, shop storage space, and system compatibility across retailers. Participants rated their likelihood of using each of three proposed reusable product types (Figure 9.4). Reusable shopping bags received the highest interest (4.2/5), though the expected adoption by others was rated lower (3.3/5). These concerns shaped the design of the prototype and motivated emphasis on usability and process transparency.

Service Concept Evaluation

The evaluation focused on how participants interpreted and experienced the proposed service, both by reviewing the service path and interacting with the return machine prototype.

Experience of the Service Path

Participants with experience of Finland's bottle return system described the service path as clear, logical, and familiar. This familiarity was widely seen as a strength, e.g., "really similar to the bottle return system" (P22). Another participant noted, "I find this service path very simple, well structured, and easy to understand" (P13). Word pair ratings supported these views, with high scores for clarity (6.0/7) and simplicity (6.1/7) (Figure 9.5).

Participants expressed strong intent to use the service (2.6/3). Many described it as a helpful fallback when forgetting to bring their own bags: "I like the fact that I can borrow these if I left mine at home" (P16). However, not all participants would use the system regularly. Some preferred personal reusable bags and questioned whether the added cost would deter use.

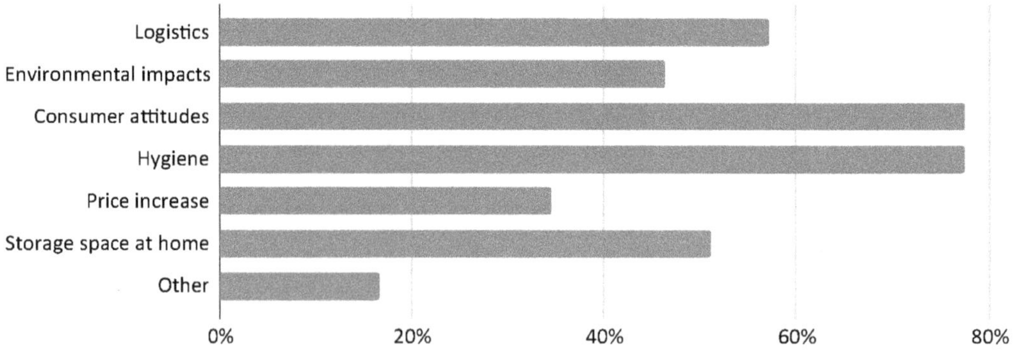

Figure 9.3 Challenges in expanding deposit-based models to other product categories.

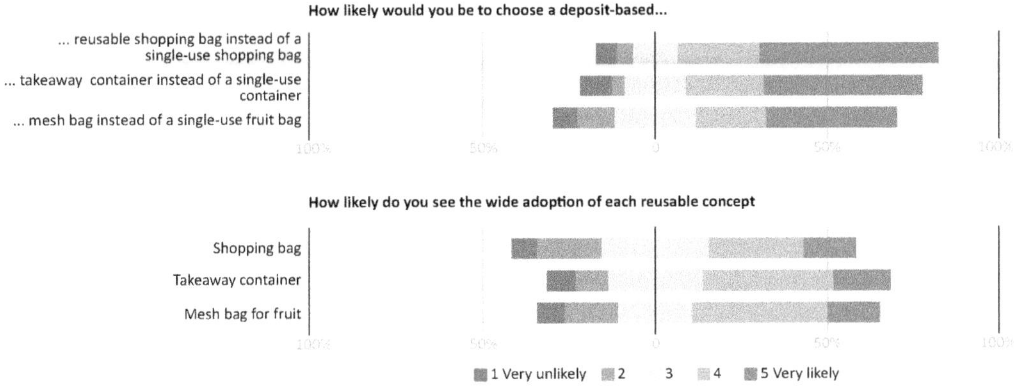

Figure 9.4 Perceived likelihood of personal use and broader adoption for reusable product concepts.

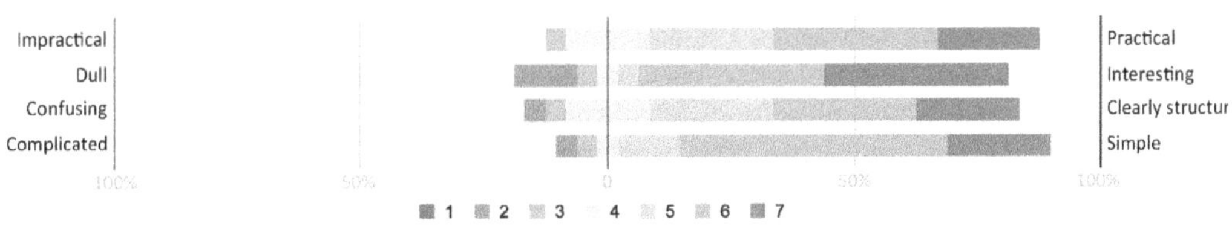

Figure 9.5 Participants' evaluation of the service path.

Participants highlighted user-side concerns such as forgetting bags, hygiene, and practical uncertainties, e.g., "What if the bag is broken? Can I still get the deposit back?" (P10). Others worried about whether unfamiliar or less tech-savvy users would find the system accessible.

From the service provider's perspective, participants raised issues related to logistics, cleanliness, and investment costs. For example: "Would it be expensive for the stores?" and "There aren't that many customers who would use it" (P17). Participants also worried about damaged or dirty returns: "People might use their bag for carrying whatever and then return it dirty… but bottles create mess also, and it's still a great system" (P24).

Interaction with the Return Machine Prototype

Interacting with the return machine prototype, participants found the system intuitive and easy to use. Its similarity to the existing bottle return machines helped create a sense of familiarity: "Reminds me of the bottle return system, easy to understand" (P7). The word pair ratings (Figure 9.6) reflected this familiarity, with strong ratings for simplicity (6.0/7), quality (6.5/7), and practicality (5.6/7). Participants also found the design attractive (5.5), creative (5.3), and even somewhat premium in feel (4.3).

Despite the positive reactions, several usability concerns emerged. The barcode scanning process was unclear to many users. Suggestions included better lighting, labels, or on-screen prompts: "Maybe instructions where the bar code reader is… could be text next to the reader too" (P11). The return slot also raised hygiene concerns, especially in public settings.

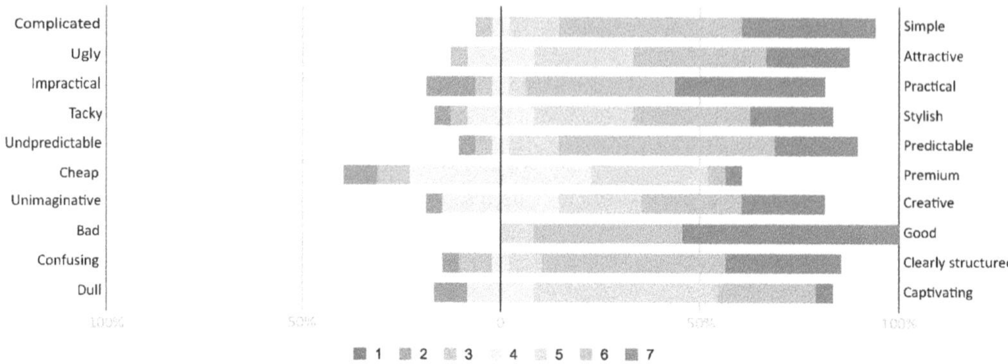

Figure 9.6 Participants' evaluation of the return machine prototype.

Participants made a range of practical suggestions:

- Replace paper receipts with digital alternatives.
- Include small rewards or positive feedback (e.g., animations or emojis).
- Clarify pricing and deposit rules for different bag types.
- Improve visual consistency across bags, machines, and signage.
- These insights highlight how familiar interaction patterns can support participation, while emphasizing the need for attention to usability, trust, and operational clarity.

Discussion

This section reflects on the study's findings in relation to engagement, adoption challenges, and the design of sustainable systems. By examining how familiar routines and physical prototyping influence participation, we consider how design can support everyday reuse practices and contribute to methodological development in participatory research.

Engagement through Familiar Interaction. The study found strong public interest in sustainable packaging, especially when systems align with known routines. Participants frequently referenced Finland's bottle return model, highlighting the importance of familiarity, simplicity, and trust. This supports prior research on how mental models and daily habits shape sustainability practices (Kuijer & Bakker, 2015; Meroni & Sangiorgi, 2011; Wever et al., 2008). Engagement, in this light, is not a matter of persuasion but a result of intuitive, repeatable interaction supported by infrastructure and design (Björgvinsson et al., 2012; Zimmerman et al., 2007).

Designing for Participation, Not Persuasion. Rather than treating engagement as a personal disposition or motivational deficit, the study positions it as a condition that emerges through interaction. Walkthroughs and physical prototypes allowed participants to imagine and evaluate their own participation, revealing how service usability and friction shape behavior. This aligns with PD traditions, where users contribute to shaping emerging systems through hands-on critique rather than abstract discussion (Björgvinsson et al., 2012; Le Dantec & DiSalvo, 2016).

Concerns and Preconditions for Adoption. Despite strong support for the concept (4.2/5), participants raised concerns about hygiene, cost, usability, and broader consumer appeal. They questioned whether stores would implement such systems without regulatory support, echoing calls for coordinated action across design, business, and policy (European Union, 2024; Tenhunen-Lunkka et al., 2024). These concerns emphasize that adoption depends not only on good design but also on clear rules, shared standards, and trust in operational feasibility.

Prototyping Everyday Sustainability. Low-fidelity experience prototyping surfaced practical issues—such as unclear scanning or discomfort with the return slot—that would likely be missed in abstract evaluations. These findings demonstrate the methodological value of using familiar, everyday artifacts in participatory inquiry. While RtD often focuses on novel technologies, this study shows its relevance for more mundane, policy-linked systems, where trust and everyday usability are key to sustainable adoption (Espinoza-Orias & Lundquist, 2025; Zimmerman et al., 2007).

Conclusion

This study explored a deposit-return service for reusable shopping bags, showing that such a concept is both feasible and positively received when grounded in well-known interaction models like Finland's bottle return system. Participants appreciated the clarity and simplicity of the service path, yet raised concerns about hygiene, inclusivity, and everyday usability. These findings reinforce the idea that successful reuse systems must minimize friction through routine-based, intuitive design—while also addressing logistical realities faced by retailers.

Beyond assessing user acceptance, this chapter contributes to broader design discussions by reframing engagement not as a personal trait or persuasive target, but as something shaped through physical interaction with accessible service infrastructures. The combination of online survey and experience prototyping enabled participants to meaningfully engage with a speculative concept and provide grounded, actionable feedback. This hands-on approach serves as a practical model for supporting sustainable behavior through service design.

Methodologically, the study demonstrates how low-fidelity, routine-based service prototypes can support participatory reflection even in mundane, policy-driven domains. RtD approaches are often applied to emerging technologies, but this work shows their value in surfacing trust, usability, and adoption barriers in everyday systems. In doing so, it expands the application of experience prototyping to align sustainable design interventions with the expectations and habits of daily life.

Acknowledgments

This study was funded by the Rethinking Plastics in a Sustainable Circular Economy (PlastLIFE) project (LIFE21-IPE-FI-PlastLIFE), co-funded by the European Union. Views and opinions expressed are, however, those of the authors only and do not necessarily reflect those of the European Union or CINEA. Neither the European Union nor the granting authority can be held responsible for them.

References

Behavioural Insights Team. (2022). A fine or a nudge? Charging for plastic bags works and is an interesting behavioural case study. Retrieved from www.bi.team/blogs/a-fine-or-a-nudge-charging-for-plastic-bags-works-and-is-an-interesting-behavioural-case-study/ (Accessed: 2025-04-23).

Björgvinsson, E., Ehn, P., & Hillgren, P.-A. (2012). Design things and design thinking: Contemporary participatory design challenges. *Design Issues*, 28 (3), 101–116.

Closed Loop Partners. (2025). Beyond the bag reports. Retrieved from www.closedlooppartners.com/beyond-the-bag/reports/ (Accessed: 2025-04-24).

Colley, A., Gröhn, W., & Häkkilä, J. (2021). Cleanleaf table: Preventing the spread of covid-19 through smart surfaces. In *Proceedings of the 20th International Conference on Mobile and Ubiquitous Multimedia (MUM 2021)*. ACM.

Ericsson, K. A., & Simon, H. A. (1993). *Protocol analysis: Verbal reports as data*. MIT Press.

Espinoza-Orias, N., & Lundquist, L. (2025). Life cycle assessment of reusable food and beverage packaging systems: A proposal of good practice. *Journal of Cleaner Production*, 145207. https://doi.org/10.1016/j.jclepro.2025.145207

European Environment Agency. (2020). Preventing plastic waste in Europe. Retrieved from www.eea.europa.eu/en/analysis/publications/preventing-plastic-waste-in-europe (Accessed: 2025-04-22).

European Union. (2015). Regulation (EU) 2015/720 amending directive 94/62/ec as regards reducing the consumption of lightweight plastic carrier bags. Retrieved from https://eur-lex.europa.eu/eli/dir/2015/720/oj (Accessed: 2025-05-06).

European Union. (2024). Regulation (EU) 2025/40 of the European parliament and of the council of 19 December 2024 on packaging and packaging waste. Retrieved from https://eur-lex.europa.eu/eli/reg/2025/40/oj (Accessed: 2025-04-23).

Finland Promotion Board. (2023). Circular economy success: Finland's recycling programme keeps bottles and cans off the streets. Retrieved from https://finland.fi/life-society/circular-economy-success-finlands-recycling-programme-keeps-bottles-and-cans-off-the-streets/ (Accessed: 2025-04-24).

Helsinki Times. (2024). Finns set new record for bottle and can returns in 2024. Retrieved from www.helsinkitimes.fi/finland/finland-news/domestic/26313-finns-set-new-record-for-bottle-and-can-returns-in-2024.html (Accessed: 2025-04-24).

Institute for European Environmental Policy (IEEP). (2022). Country fact sheet: Finland – deposit refund scheme. Retrieved from https://ieep.eu/wp-content/uploads/2022/12/FI-Deposit-Refund-Scheme-final.pdf (Accessed: 2025-04-24).

Jakovcevic, A., Steg, L., Mazzeo, N., Caballero, R., Franco, P., Putrino, N., & Favara, J. (2014). Charges for plastic bags: Motivational and behavioral effects. *Journal of Environmental Psychology*, 40, 372–380.

Kuijer, L., & Bakker, C. (2015). Of chalk and cheese: Behaviour change and practice theory in sustainable design. *International Journal of Sustainable Engineering*, 8 (3), 219–230.

Le Dantec, C. A., & DiSalvo, C. (2016a). Infrastructuring as a practice of design. *Journal of Human-Computer Interaction*, 30 (6), 534–555.

Manzini, E. (2015). *Design, when everybody designs: An introduction to design for social innovation*. MIT Press.

Meroni, A., & Sangiorgi, D. (2011). *Design for services*. Gower Publishing, Ltd.

Miao, X., Magnier, L., & Mugge, R. (2023). Switching to reuse? An exploration of consumers' perceptions and behaviour towards reusable packaging systems. *Resources, Conservation and Recycling*, 193, 106972.

Pitkänen, K., Peltola, T., Onali, A., Heikkilä, W., & Fjäder, P. (2024). Miten vähenämme muovikassien kulutusta? kansalaispoolin ratkaisut ja suositukset. plastlife. Retrieved from https://kiertotalousratkaisuja.syke.fi/kansalaispoolin-suositukset-1-2024-miten-vahennamme-muovikassien-kulutusta/ (Accessed: 2025-05-06).

Stickdorn, M., Lawrence, A., Hormess, M. E., & Schneider, J. (2018). *This is service design doing: Applying service design thinking in the real world*. O'Reilly Media.

Tenhunen-Lunkka, A., Balatsas-Lekkas, A., Mouazan, E., Palola, S., Ngo, T., Salo, M., … & Sorvari, K. (2024). Implementing a circular business model for reusable packaging: Multidisciplinary learnings from reusable pizza packaging. *Sustainable Production and Consumption*, 48, 62–83.

United Nations Environment Programme. (2018). *Single-use plastics: A roadmap for sustainability*. UNEP. Retrieved from www.unep.org/resources/report/single-use-plastics-roadmap-sustainability

Wever, R., Van Kuijk, J., & Boks, C. (2008). User-centred design for sustainable behaviour. *International Journal of Sustainable Engineering*, 1 (1), 9–20.

Ympäristöministeriö. (2023). *Muovikassien vähentämisen green deal -sopimuksen arviointi 2023 (Tech. Rep.)*. Ympäristöministeriö.

Zimmerman, J., Forlizzi, J., & Evenson, S. (2007). Research through design as a method for interaction design research in HCI. In *Proceedings of the SIGCHI Conference on Human Factors in Computing Systems*, 493–502. ACM.

10 Designing for Green Transitions

A Case Study of Reusable Takeaway Bowls in Airports

Elisabeth Tschavgova, Sonja van Dam, Conny Bakker, Jo van Engelen, and Rita Jonyer

Introduction

Given the urgency of the climate crisis, the European Commission has identified the circular economy (CE) as a key priority in achieving sustainability (European Commission, 2018). Transitioning to a CE is seen as an effective strategy for reducing environmental impacts across various sectors (Pizzi et al., 2022), including aviation. In response, Norwegian airports, part of the Avinor group, are working towards achieving circularity (Avinor, 2023). For instance, Oslo Airport aims to halve its residual waste from daily operations by 2030 through various zero-waste projects (Avinor, 2023). These ambitions are not only driven by internal goals but also by compliance with external regulations, such as the European Green Deal. A key regulation in this transition is the EU Packaging and Packaging Waste Regulation (PPWR), which aims to minimise packaging waste and reduce reliance on primary raw materials. In alignment with this, Oslo Airport is exploring the adoption of reusable takeaway packaging through a pilot of reusable bowls combined with emerging technologies for payment and smart return machines. Based on this pilot, the study aims to explore driving green transitions within a complex organisational setting using participatory engagement. The following research questions guide this study:

RQ1: "What are the key enablers and obstacles to employee and passenger engagement in the case study, taking operational feasibility into account?"
RQ2: "What broader learnings on interventions advancing green transitions in complex environments, such as an airport, can be drawn from this case study?"

This research contributes to the broader field of sustainability transitions by addressing the limited empirical studies on the implementation of reusable takeaway packaging systems (Bradley & Corsini, 2023). In doing so, it also fills a gap in the operationalisation of CE principles within the airport contexts (Van Der Tuin-Rademaker et al., 2024). By following a systemic design (SD) approach based on combining participatory action research and a causal-loop diagram (CLD) to synthesise findings, this chapter offers insights into how system dynamics, particularly consumer and employee engagement, influence circular transitions in complex environments.

The chapter is organised as follows: First, a literature review of academic contributions on the adoption of CE principles in airports and the implementation of reusable packaging to position this study is conducted. Next, the methodology of the pilot is introduced, followed by the synthesised findings presented in the CLD that highlight insights regarding passenger

DOI: 10.4324/9781003667872-13

and employee engagement. Finally, the discussion reflects on the pilot and broader learnings, addressing the challenges and opportunities associated with adopting reusable packaging.

Literature Review

Airports encounter unique challenges in adopting sustainable practices. Therefore, this literature review explores the complexities of implementing reusable packaging systems in these environments and identifies empirical gaps in current research. By examining two critical themes – **airports as complex environments for implementing CE principles** and **existing research gaps regarding reusable packaging** – this study aims to establish its theoretical foundation.

Airports as Complex Environments for Implementing CE Principles

As circular transitions require a systemic approach (Guzzo et al., 2022), airports function as complex systems characterised by interconnected relationships among various actors, technologies, and processes (Aarikka-Stenroos et al., 2021). However, stakeholders often operate in silos, leading to inefficiencies across the airport system (Parra, 2023) and limited adoption of CE concepts due to diverse stakeholder interests (Van Der Tuin-Rademaker et al., 2024). Stakeholder management is further complicated as their goals may not always align with those of the airport management (Schaar & Sherry, 2010). Thus, collaboration and effective resource management are fundamental for a CE transition (Sassanelli & Terzi, 2023), as is the engagement of a range of stakeholders, including employees, passengers, and industry partners (Van Der Tuin-Rademaker et al., 2024). While the literature found contributes to advancing CE principles in aviation by discussing obstacles and opportunities, much of the research remains theoretical, with limited description of real-life interventions. Studies such as those by Tjahjono et al. (2024), Baxter (2022), and Özbay and Gokceviz (2022) have laid out conceptual frameworks and models for waste management and CE transitions in airport settings. However, these studies do not explore the practical implementation of their suggested approaches, focusing instead on theoretical constructs and potential strategies without real-world application or case study evidence. This gap underscores the need for more empirical research to translate theoretical suggestions into actionable interventions within the complexity of the airport environment. Thus, to navigate the airport complexities and their system actors to develop a feasible intervention, an SD approach, which integrates systems thinking and design (Bijl-Brouwer & Malcolm, 2020), offers a promising method. Due to its co-creative nature (Jones, 2018), SD provides a promising approach to facilitate systemic interventions through engagement, which is crucial for diverse organisational activities such as value creation, innovation, and sustainability (Kujala et al., 2022), while collaboratively addressing challenging issues (Oberholzer & Sachs, 2023). Therefore, this study defines engagement as the participatory collaboration of various stakeholders – employees, passengers, and industry partners – in the development of SD interventions to advance green transitions.

Empirical Gaps in Reusable Packaging Research

A recent literature review by Bradley and Corsini (2023) reveals that most research on reusable packaging relies on quantitative methods, such as life cycle assessment, which, while providing measurable indicators of success, lacks empirical validation through case studies. This aligns with Tenhunen-Lunkka et al. (2024), who noted the scarcity of empirical field studies and limited research on consumer attitudes toward reuse. Ertz et al. (2017) further highlights the absence of consumer perspectives on reusable packaging as a significant oversight. In addition to a stronger focus on consumer attitudes, research on reusable takeaway packaging requires interdisciplinary collaboration among various stakeholders (Pålsson & Olsson, 2023). Bradley and Corsini (2023) emphasise the need for multidisciplinary perspectives, as current research is predominantly technical and neglects social sciences, user experience, and service design. Despite the recognised importance of stakeholder collaboration (Chan, 2007), existing research remains fragmented across fields. Therefore, to advance the field, a comprehensive multi-stakeholder approach, involving retailers, manufacturers, logistics companies, policymakers, and consumers (Bradley & Corsini, 2023) is needed. The reviewed literature suggests an empirical gap in the implementation of reusable packaging, particularly a need for studies that adopt a multidisciplinary perspective with a focus on design.

To ensure that no relevant case studies or empirical research that could offer insights into the adoption of reusable packaging alternatives within airport environments were overlooked, a search string based on the six key concepts of this study was developed:

("Circular Economy" OR "Sustainable practices" OR "Green transitions")
AND
("Airports" OR "Aviation industry" OR "Airport management")
AND
("Engagement" OR "Participatory design" OR "Stakeholder collaboration")
AND
("Reusable Packaging" OR "Sustainable packaging" OR "Reusable containers" OR "Packaging waste reduction")
AND
("Case Study" OR "Case Studies" OR "Empirical research")
AND
("Systemic Design" OR "Systemic approach" OR "Systems thinking" OR "Systemic intervention")

When applying this search string with a preference for English articles and no restriction on the time period, very limited results were found. Databases such as Web of Science and Scopus indicated no matching literature, and Google Scholar suggested 109 results (excluding citations). Upon reviewing the titles and abstracts of found articles, no literature was identified that constituted a case study adopting reusable packaging within an airport context. This lack of matching case studies indicates a gap concerning reusable packaging within airports. It underscores the need for more empirical studies on real-world applications of reusable packaging systems in complex environments. Therefore, the limited case studies reinforce this research's empirical aim to achieve a systematic and design-oriented perspective. To the best of our knowledge, this chapter represents one of the first efforts to fill in the gap in empirical implementation of reusable packaging in complex environments like airports, while integrating multi-stakeholder perspectives.

Methodology

The research employs a case study following an action research approach to assess the impact of reusable takeaway packaging at the airport and identify key engagement enablers and obstacles. The case study follows the plan-act-observe–reflect cycle of action research (Interaction Design Foundation [IxDF], 2015), as shown in Figure 10.1.

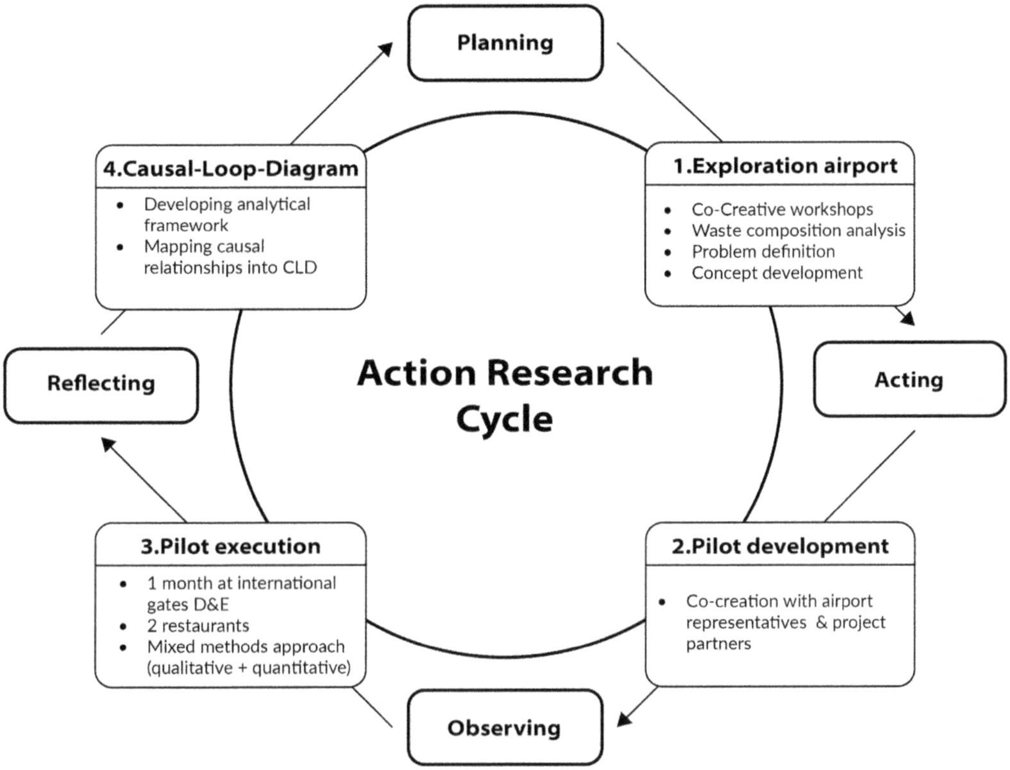

Figure 10.1 Action research process embedded in case study.

Source: Adapted from Interaction Design Foundation (IxDF), 2015.

Exploration Airport

To gain insights into the current airport waste management system, co-creative workshops were conducted with various airport representatives, and a waste composition analysis was performed, including mapping material flow and examining resource stream compositions based on weight, volume, and treatment methods, while also verifying the treatment processes (Van Der Tuin-Rademaker et al., 2024). These activities provided a comprehensive overview of the waste generated as well as insights into the composition of the residual waste stream. This stream represents the largest portion of waste generated by internal processes and currently goes straight into incineration for energy recovery. Several scoping sessions and student projects were conducted to develop sustainable interventions aimed at reducing residual waste. One project focused on dishwashing-as-a-service as an alternative to single-use takeaway. This led to the reusable takeaway packaging system "Green Circles" (Falcón, 2024).

Pilot Development

To test "Green Circles", a single case study approach was adopted as defined by Yin (2014). This case study was selected for its ability to capture real-world interactions, identify systemic barriers, and provide contextually grounded insights that can inform both academic discourse and practical decision-making regarding CE transitions.

The case study was conducted at Oslo Airport, with the objective of testing a fully operational infrastructure for reusable packaging. The pilot was conducted over the course of one month at the airport's international D and E gates and involved two participating restaurants. The development of this infrastructure was carried out iteratively, in collaboration with Avinor (including the circularity team, restaurants, and cleaning facilities) and external stakeholders (smart return machine provider, payment system provider, and reusable bowl provider). The pilot featured QR-coded reusable bowls integrated with a point-of-sale (POS) system, smart return machines, a centralised dishwashing service, and promotional materials (Figure 10.2). The design of the promotional materials and the machines was developed collaboratively, ensuring consistency with Avinor's brand style. The promotional materials consisted of A5 and A4 posters, which adhered to the same visual language as the machines. The posters were placed in table stands, with one side in English and the other in Norwegian, and were placed on all tables and next to the registers of both restaurants. The posters provided a brief visual explanation of the reusable bowl and a QR code link to a subpage on avinor.no for further details. Online promotion materials also extended to internal communication channels for employees and social media posts on LinkedIn.

Customers who purchased food from these restaurants could opt for a reusable bowl, equipped with a unique QR code for traceability. At checkout, customers tapped their payment cards at the POS, which followed a novel "tap-and-reuse" approach. When a customer tapped their card, no deposit was withdrawn from their bank account. They had one week to return the bowl to one of the six designated smart return machines at the international gates D and E. Only if a customer failed to return the bowl within this timeframe was a fee of 70 NOK automatically deducted from their account. Returned bowls were collected from the smart return machines by airport cleaning staff and transported to the central dishwashing facility within the airport, where they were cleaned and redistributed to the participating restaurants (Figure 10.3).

Pilot Execution

To capture the various perspectives and realities of transitions within a complex environment, this study utilised a mixed-method approach, combining both quantitative and qualitative data sources. Surveys that focused on passengers' perceptions of reusable packaging (n = 59) provided quantitative data. These surveys were conducted by engaging passengers after they

Figure 10.2 (a) Promotion material and reusable bowl; (b) Smart return machine; (c) Scanning QR code of reusable bowl by employee to activate POS; (d) Customer tapping with card at POS to activate the deposit-free borrowing system.

Source: Photos @Rita Jonyer.

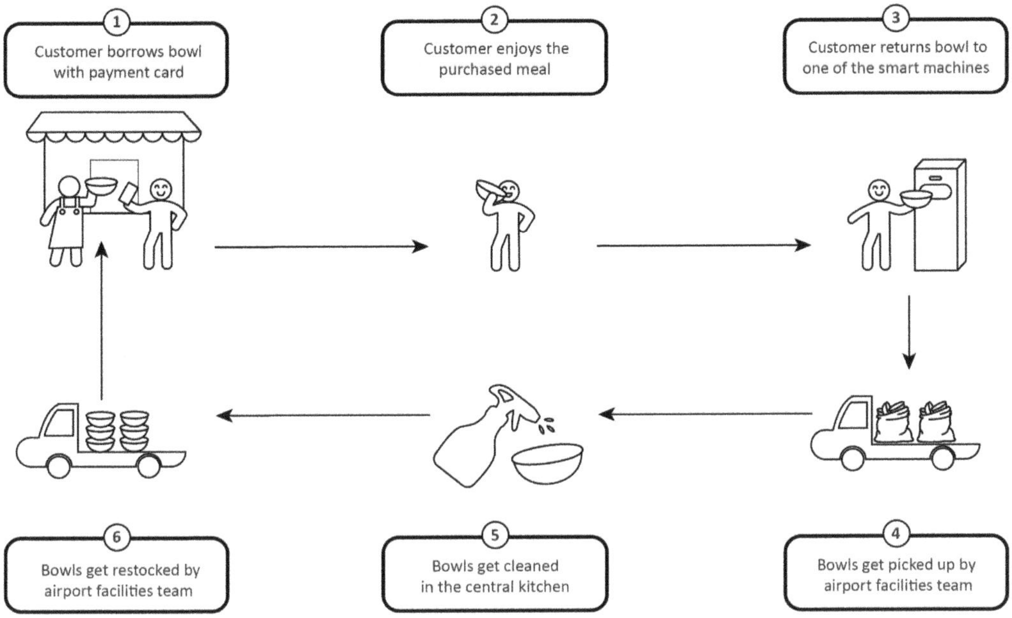

Figure 10.3 Process of the pilot: Steps 1–3 show the front end of the bowl borrowing process, and steps 4–7 show the backend in terms of pick up, cleaning, and redistribution of the bowls.

purchased a takeaway meal at one of the two participating restaurants. Participants were asked where they planned to eat their meal, whether they were aware of a reusable packaging alternative, and if so, why they were (not) choosing it. Qualitative data was collected through various means, including interviews ($n = 9$) with a diverse group of system actors, such as employees, project partners, and industry experts. These interviews offered a multi-stakeholder perspective on engagement with reusable takeaway packaging. Furthermore, on-site observations (50 hours over one month) were conducted at key locations, such as restaurants, smart return machines, and seating areas. Informal conversations were conducted with employees, passengers, and other airport staff interested in the reusable packaging. A field diary (Couvinhas et al., 2015) was kept throughout the pilot, documenting observations, insights, and informal conversations with various system actors. The survey questions and semi-structured interview guides can be found online in appendix A (full appendix link supplied at the end of the chapter).

Causal-Loop-Diagram

The final phase of the research involved synthesising the data collected to extract key insights, identify patterns, and develop a conceptual model illustrating the dynamics influencing the transition to reusable takeaway packaging in the complex environment of an airport. An analysis framework based on the research questions was created to structure and guide the analysis, ensuring alignment between the study's objectives and emergent themes. The framework (Figure 10.4) revolves around eight key themes. The four themes of RQ1 were recurring topics throughout the entire pilot, raised by internal and external stakeholders. "Communication and information flow" and "consumer engagement" were frequently discussed with the internal

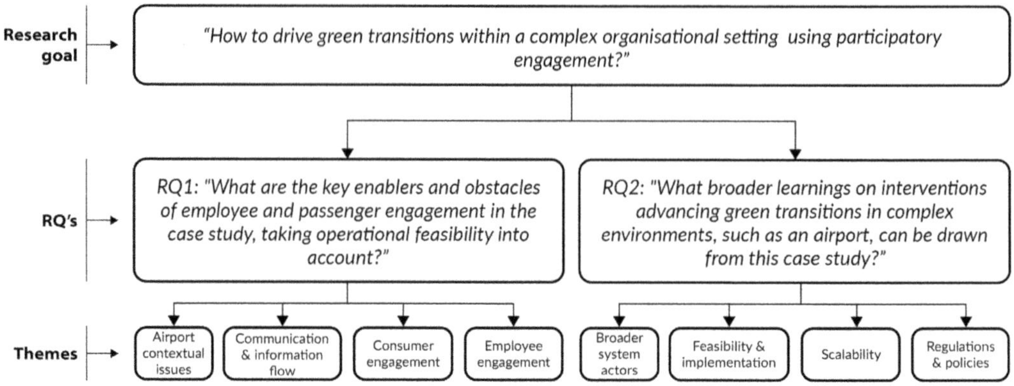

Figure 10.4 Conceptual framework based on the thematic analysis.

stakeholders during the pilot preparations to ensure that the system was functioning and viable. Discussions with external stakeholders, who provided the return machines and POS, addressed "airport contextual issues" and emphasised "employee engagement" to promote the reusable bowls. The themes "Broader System Actors" and "Regulations & Policies" of RQ2 emerged from interviews with industry experts as key enablers for promoting reusable packaging. Subsequent coding and clustering of relevant quotes led to the final topics: "Feasibility & Implementation" and "Scalability".

The interview transcripts, relevant observational notes, and researcher reflections were digitised and imported into Atlas.ti. The textual insights were systematically coded, and the eight key themes were refined through iteration. To go beyond a siloed representation of individual themes, a CLD was developed to visualise the recurring patterns and causal relationships between the diverse insights (Bala et al., 2017). While the analytical framework helped categorise the findings thematically, the CLD reveals how these themes influence one another through feedback loops and interdependencies.

Results

The results present the case study findings, structured to address the two research questions through the lens of the eight analytical themes. First, findings on the airport's organisational context are described to establish the spatial, behavioural, and operational characteristics that make it a uniquely challenging environment for reusable bowls. This contextual understanding of the airport provides the foundation for many of the barriers and enablers discussed throughout this chapter. Next, the results from the pilot execution are presented, particularly the limited borrowing rate and the interdependence of employee and consumer engagement. These findings are essential to understanding how certain behaviours and operational routines hindered the implementation of the reusable bowl. These results also serve as the basis for several of the eight analytical themes, particularly those related to engagement, feasibility, and contextual barriers. Finally, the CLD is introduced, which embeds the dynamic interrelations across all eight themes.

Defining the Airport Environment

The case study highlights a unique challenge for implementing reusable packaging in the airport environment with its transient, high-pressure nature. Unlike traditional food service locations, where customers return frequently and can gradually establish new habits, airports cater to a constantly changing flow of passengers. Each traveller must learn about reusable packaging from scratch, yet many leave without returning, making it a semi-open-loop environment with difficulties in creating lasting behavioural change. The stressful nature of air travel further complicates engagement. Passengers are often preoccupied with boarding procedures, navigating security, and reaching their gates on time, leaving little room for learning and adopting a new behaviour. This fast-paced environment also puts pressure on employees, who face stressful passenger peak hours, where explaining the reusable packaging adds to the already intense workload. The international nature of the airport also introduces language and cultural barriers, adding another layer of complexity to the adoption of a reusable system. Overall, various unique airport-specific constraints became evident when integrating reusable takeaway packaging during the pilot (Figure 10.5).

Pilot Execution

The pilot resulted in the low borrowing rate of 35 reusable bowls within one month. The average return time for a bowl varied, with passengers returning them on the same day, while some airport employees keeping the bowl for up to five working days. Additionally, the pilot faced several technical challenges, such as internet connectivity, POS device malfunction, currency errors and smart return machines being disconnected by passengers.

The primary reason for the low turnout was that most of the takeaway food was pre-packaged in disposable packaging, leaving passengers with no reusable alternatives. Of the 59 passengers surveyed after purchasing meals at the participating restaurants, 44% received their meals prepackaged in disposable packaging, making the availability of reusable options irrelevant in those cases. Further 49% reported having no awareness of the reusable packaging option. Interviews with employees corroborated the findings as the employees were not actively promoting the reusable bowl due to language barriers, other work priorities, additional workload needed to explain reusable packaging during peak hours, not feeling informed enough or established work routines defaulting to single use. During informal conversations, passengers expressed limited interest in using a new system because they felt stressed about navigating the unfamiliar airport environment, had little time before boarding, and viewed the system as potentially risky due to uncertainty about locating a smart return machine in time. Consequently, employees also felt demotivated.

Observations revealed that disposable packaging as the default was due to prepacked meals being served in disposable packaging as standard. This was partly a result of logistical processes in airport food production and partly due to employees not asking passengers about their intended consumption location and defaulting to disposables. However, the passenger survey indicated that a majority of 66% planned to eat at the restaurant, 15% at their gate or another terminal location, 14% intended to take their food onto the plane, and 5% planned to eat in other locations. Ultimately, the lack of passenger awareness and employee routines defaulting to disposable packaging were identified as key bottlenecks in the pilot's low turnout.

Constraints of the airport environment	
Limited space in airport restaurants	Limited space makes it a challenge to install additional smart return machines and payment devices, as well as to store reusable bowls at the restaurants in addition to the disposable ones
Time constraints	Reusable packaging preferably needs to be returned before a flight departs
Airport restaurants have different ways of working	A standardised approach was difficult to apply across all serving units
International payment requirements	Multiple card payment options for borrowing reusable packaging and processing 'authorisation holds' were required and posed some technical challenges
Businesses' pricing policies	Some restaurants had pricing policies that made takeaway meals cheaper than dine-in options, indirectly incentivising disposable packaging
Pre-packaged food practices	Some food items arrived at restaurants already prepacked at a centralised kitchen in disposable packaging, limiting the scope of reuse
Security regulations and infrastructure challenges	• Restricted internet accessibility is affecting real-time data tracking and payment processing. • Limited access for external personnel is making installation and troubleshooting more difficult for external stakeholders. • Pre-approval requirements for any infrastructure changes, such as installing smart return machines or modifying payment systems, significantly slowed down the implementation process
Unpredictable passenger and employee behaviour	• Passengers unplugging smart machines to charge their phones, disrupting operations. • Passengers misidentifying the machine as luggage trollies' return stations due to airport branding, blocking the machine's usage. • The disappearance of cables and phone stands, potentially due to theft, requiring unexpected replacements

Figure 10.5 Airport environment constraints.

Causal-Loop-Diagram

To synthesise the findings and visualise the complex relations of the 8 themes of the analytical framework, a CLD (Figure 10.6) was developed. The diagram visually maps key feedback loops and interdependencies between the themes that emerged from the thematic analysis, providing an integrated systems-level understanding on how reusable takeaway packaging interventions operate in the complex setting of an airport. The CLD addresses both research questions by organising insights into two overarching clusters: **Case-Specific Engagement and Feasibility** (RQ1) and **Transition Enablers** (RQ2). The embedded themes in the CLD are briefly elaborated below, and detailed summaries of the themes can be found online in appendix B.

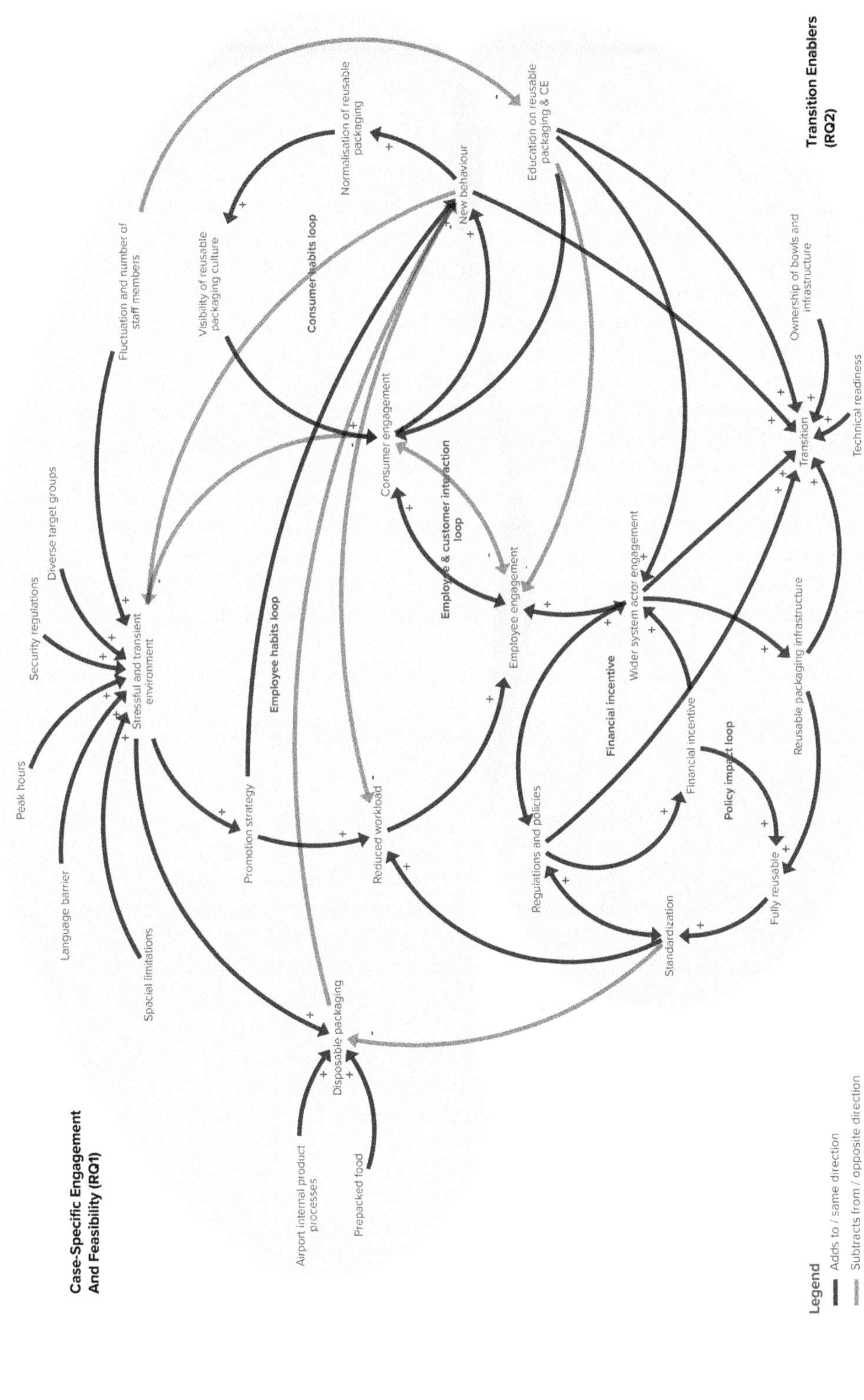

Figure 10.6 Causal-loop-diagram (high resolution available online in appendix C).

Case-Specific Engagement and Feasibility (RQ1)

Airport contextual issues: The CLD shows the complexity of the airport environment, highlighting unique barriers such as spatial limitations, language barriers, diversity in target groups and time pressure. These barriers contribute to the transient and stressful nature of the airport's open-loop system, putting it in a less suited position to lead the transition to reusable packaging.

Communication and information flow: Clear and consistent communication was vital for employee and passenger engagement during the pilot. However, the airport environment presented challenges, such as fluctuating staff numbers over time and an overall high number of staff (20–60 per restaurant). This made it difficult to ensure that all employees were well-informed. Increased customer awareness regarding reusable bowls, additional staff training, and accessible information were suggested for broader engagement with the system.

Entanglement of "Consumer engagement" and "Employee engagement": The CLD reflects the interdependence between employee engagement and consumer uptake. As seen in the pilot, without active promotion from employees, the system remained largely invisible to passengers. The variables "Visibility of reusable packaging culture" and "Normalisation of reusable packaging" in the consumer habits loop illustrates the reinforcing relationship needed to establish a new consumer behaviour and its relation to employee engagement. When reusable packaging becomes more visible, passengers are more likely to adopt it. As adoption grows, the behaviour becomes self-reinforcing, reducing the need for active employee promotion and lowering their workload. This, in turn, enhances the overall passenger experience and boosts engagement among both employees and passengers. Additionally, these loops highlight the significance of clear communication and education in facilitating engagement. Enhancing promotion, simplifying touchpoints, and integrating system information into staff training and ads can help reduce friction and encourage participation.

Transition Enablers (RQ2)

Broader system actors: Engaging both internal and external stakeholders proved essential for aligning incentives, standardising infrastructure, and establishing shared ownership of the reusable packaging system. This becomes even more relevant when scaling beyond the airport, as no single actor can drive this transition alone. Instead, collective action from a consortium of brands, retailers, and public bodies is necessary. Additionally, the CLD indicated a positive impact on the engagement of diverse system actors through education on CE and the benefits of reusable packaging.

Feasibility & implementation: The pilot revealed that implementing reusable packaging in an airport context is constrained by lack of engagement and internal logistics. Additional challenges stemmed from established operational routines, such as the use of pre-packed food in disposable packaging, as well as technical issues with the emerging technologies utilised during the pilot. These findings highlight the importance of robust infrastructure, regulatory alignment, and a context-specific design. These dynamics are captured across various loops and variables within the CLD, as the operational feasibility insights are spread throughout.

Scalability: Scalability of reusable packaging depends on standardisation, regulation, and ownership. Standardization across product types, infrastructure, and return processes reduces friction and enables broader adoption, while granting individual vendors flexibility. National and EU regulatory frameworks can position reuse as the default option, enhancing its financial viability and creating a self-reinforcing loop of standardisation and scalability. Ownership needs to align with system scale, while airports may manage infrastructure locally, broader expansion requires government or consortium-led models.

Policy & regulations: In the CLD, two primary loops emerged concerning this theme. The "Policy impact loop" focuses on how policies encourage infrastructure development and standardisation, making reusable packaging feasible for businesses to adopt. The second loop, "Financial incentive", illustrates how regulations associated with financial penalties or rewards can motivate wider system stakeholders to prioritise and enforce reuse.

Discussion

This research explores how participatory engagement and SD approaches can drive green transitions in complex organisational settings, illustrated by a case study on reusable takeaway bowls. The discussion reflects on the eight themes identified in the CLD, their broader implications, theoretical contributions, and study limitations.

Contextual Factors Shaping Sustainable Behaviour

This case study supports existing literature on how contextual factors, such as time pressure and environmental unfamiliarity, influence sustainable behaviour in airports, affecting both passenger and employee engagement (Ertz et al., 2017). For instance, environmentally conscious employees faced challenges promoting reusable bowls due to peak-hour pressures, limited system knowledge, and stressed passengers. This stress aligns with literature on perceived time scarcity among consumers (Ertz et al., 2017), potentially heightened in airport settings. The concept of prosumption (Ritzer & Jurgenson, 2010) is relevant here, as consumers increasingly take on tasks once performed by employees, such as check-ins and digital ordering, shifting responsibility and cognitive load onto themselves. This shift increases the desire for speed and convenience, making disposable packaging more appealing, while reusable alternatives may seem like additional tasks. Thus, reusable systems must require minimal cognitive effort and integrate seamlessly into existing service flows, especially in high-stress environments like airports. These findings underscore that intrapersonal variables, such as intentions or environmental values, are insufficient for sustainable behaviour (Ertz et al., 2017; Steg & Vlek, 2009). Contextual "nudges" are more effective than those based solely on value orientation or information provision (Campbell-Arvai et al., 2014). This was evident in the pilot, where promotional campaigns and employee ambassadors alone were insufficient. Insights suggest potential for context-based interventions, like green procurement policies that charge extra for takeaway to discourage disposable containers. However, some barriers, such as work culture, limited space, diverse payment systems, and cultural differences due to international backgrounds, are inherent to airports and not easily solvable.

Theory of Planned Behaviour and Engagement at the Airport

The insights align with the Theory of Planned Behaviour (TPB), which frames behavioural intent as being influenced by perceived control, subjective norms, and attitudes (Ajzen, 1991). The pilot demonstrated that behavioural intent may not predict behaviour when 'actual behavioural control', including environmental constraints and facilitating conditions, is lacking. For instance, employees with positive sustainable attitudes struggled to promote the bowls due to high workloads or pre-packaged meals. Similarly, passengers' use of reusable bowls was hindered by weak subjective norms, time constraints, the unknown airport environment, and insufficient promotions. The CLD further aligns with TPB by identifying factors that hinder behavioural change. Passengers not seeing others return the bowls weakened subjective norms, and those with pro-environmental attitudes were discouraged by stress or uncertainty about the return process. Employees faced technical issues or were unable to promote reusable bowls during peak hours, limiting their perceived control. Collectively, these barriers undermined engagement.

Strategies and Enablers

The pilot revealed both challenges and feasible strategies within the airport environment. Reusable packaging systems must be designed to be visible, convenient, and frictionless while considering context constraints to enable normalisation. Simultaneously, disposable alternatives should be made less accessible or attractive. As Ertz et al. (2017) argue, environments that facilitate reuse and discourage disposables shift perceived utility toward sustainable choices. When passengers perceive reusables as the default or less of a hassle, they are more motivated to choose them. Future research could implement a phased approach to promote reusable bowls, starting with employees to establish internal norms, then extending to domestic flights familiar with reuse practices. Once integrated, additional reusable packaging can be introduced, leading to full airport-wide implementation. In addition, stricter national regulations and supportive airport policies can accelerate the transition to reusable packaging. However, questions of ownership arise as infrastructure grows, requiring effective governance structures with clear accountability and lifecycle oversight.

Interventions Required Beyond the Techno-fix to Drive Green Transitions

The pilot offers broader insights into interventions advancing green transitions. While emerging technologies like smart return machines and QR-linked POS devices are essential for operationalising reusable packaging, they alone are insufficient for systemic change. Drawing on Sætra and Selinger (2024), the reusable packaging system exemplifies a techno-fix – "*a technological remedy for a social problem involving minimal political or economic incentives, without requiring changes in social norms.*" Framing reusable packaging as a standalone solution risks ignoring necessary socio-political and institutional changes. The pilot found that airport practices, such as prepacking food in disposable packages or restaurant staff not inquiring about passengers' consumption preferences, are as influential as technical setups in waste reduction. These findings emphasise that beyond reusable infrastructure, behavioural adjustments and policies are crucial. Therefore, both top-down interventions, which reshape market conditions through legislation and incentives, and bottom-up initiatives, which drive innovation through new technologies and business models, are essential (Guzzo et al., 2022). The CLD confirms that supportive regulation plays an enabling role in changing consumer behaviour towards reusable packaging and further indicates that an airport alone cannot lead such a transition. Broader regulations, such as the PPWR, create top-down pressures and opportunities for standardisation, aligning efforts across the value chain and normalising circular practices. This requires collaboration at all levels from diverse system actors, as found in the "Transition Enablers" cluster of the CLD, aligning with Fichter et al. (2023).

SD Addressing Complex Sustainability Challenges

The pilot demonstrated the value of SD in navigating the complexity of highly regulated, multi-stakeholder environments like the airports. As Jones (2014) argues, SD is well-suited for addressing "wicked problems" in sustainability, where actors, goals, and systems co-evolve. In this study, SD enabled iterative, co-creative processes involving various airport departments, hierarchical levels, and external stakeholders to collaboratively design systemic interventions. The SD approach further aligns with critiques by Sætra and Selinger (2024), who highlight that framing societal issues as purely technical can obscure their political, cultural, and human dimensions. Following an SD approach reframes technology as a tool for deeper socio-behavioural change rather than the solution itself. Thus, following the holistic and co-creative SD approach, the risk of reusable packaging becoming a techno-solutionist endeavour that overlooks the constraints of the complex contextual challenges, internal processes, and regulations can be reduced.

Limitations

This study provides context-specific insights into reusable packaging adoption within an airport environment with several limitations. First, the findings are based on a single pilot, which limits the insights generalisability. More case studies across different contexts are needed to validate and generalise the findings. Second, the environmental impact of the reusable bowl was not assessed through a life cycle assessment, which could offer insights into sustainability regarding emissions, water usage, and material consumption. Lastly, the pilot was brief and limited in scope, focusing on one intervention model. Broader engagement, longer timeframes, and varied contexts are needed to test the long-term feasibility and adaptability of reusable take-away packaging.

Conclusion

This case study explores the implementation of a reusable bowl system in the complex environment of an international airport using an SD approach. The pilot revealed that technological interventions alone are insufficient for driving sustainable transitions. Behavioural, contextual, and institutional factors also play critical roles. At the case-specific level, the findings highlight the interdependence of passenger and employee engagement, both heavily influenced by the airport environment. Airport-specific constraints such as stress, time scarcity, spatial limitations, and cultural diversity posed challenges to implementing reusable bowls, suggesting that scaling such solutions would be easier in less demanding environments. Nonetheless, the pilot identified feasible interventions like green procurement, phased implementation, and increased promotion. Beyond the pilot, insights emphasise the need for alignment between top-down and bottom-up efforts, such as government incentives for reusable packaging and startups developing new technologies. The CLD from this study visualises the complexity of transitioning to reusable packaging, demonstrating that techno-fixes alone are insufficient. Social and institutional transformations are essential for advancing green transitions.

Acknowledgement

This study was part of the EU project TULIPS (2022) which received funding from the European Union's Horizon 2020 research and innovation program under grant agreement No. 101036996.

References

Aarikka-Stenroos, L., Ritala, P., & Thomas, D. W., L. (2021). Circular economy ecosystems: A typology, definitions, and implications. In S. Teerikangas, T. Onkila, K. Koistinen, & M. Mäkelä (Eds.), *Research Handbook of Sustainability Agency* (pp. 260–261). Edward Elgar Publishing. https://doi.org/10.4337/9781789906035.00024

Ajzen, I. (1991). The theory of planned behavior. *Organizational Behavior and Human Decision Processes, 50*(2), 179–211. https://doi.org/10.1016/0749-5978(91)90020-T

Avinor. (2023). Annual and sustainability report. https://avinor.no/contentassets/b5d94158f9de40709e917343fde524aa/avinor_arsrapport_2023-eng.pdf

Bala, B. K., Arshad, F. M., & Noh, K. M. (2017). Causal loop diagrams. In B. K. Bala, F. M. Arshad, & K. M. Noh (Eds.), *System Dynamics: Modelling and Simulation* (pp. 37–51). Springer. https://doi.org/10.1007/978-981-10-2045-2_3

Baxter, G. (2022). Towards sustainable airport waste management through the adoption of a "green" airport strategy: The case of Incheon International Airport. *Clean Technologies and Recycling, 2*(4), 247–278. https://doi.org/10.3934/ctr.2022013

Bradley, C. G., & Corsini, L. (2023). A literature review and analytical framework of the sustainability of reusable packaging. *Sustainable Production and Consumption, 37*, 126–141. https://doi.org/10.1016/j.spc.2023.02.009

Campbell-Arvai, V., Arvai, J., & Kalof, L. (2014). Motivating sustainable food choices: The role of nudges, value orientation, and information provision. *Environment and Behavior, 46*(4), 453–475. https://doi.org/10.1177/0013916512469099

Chan, H. K. (2007). A pro-active and collaborative approach to reverse logistics—a case study. *Production Planning and Control, 18*(4), 350–360. https://doi.org/10.1080/09537280701318736.

Couvinhas, A. F., Paul, B., Coelho, D. A., Ferrara P., & Walter J. (2015). The field diaries: An innovative participatory methodology. In *Proceedings 19th Triennial Congress of the IEA* (Vol. 9, p. 14).

Ertz, M., Huang, R., Jo, M.-S., Karakas, F., & Sarigöllü, E. (2017). From single-use to multi-use: Study of consumers' behavior toward consumption of reusable containers. *Journal of Environmental Management, 193*, 334–344. https://doi.org/10.1016/j.jenvman.2017.01.060

European Commission. (2018). *A Clean Planet for All, A European Strategic Long-Term Vision for a Prosperous, Modern, Competitive and Climate Neutral Economy.* European Commission, Brussels.

Falcón, M. G. (2024). Designing dishwashing as a service at Oslo Airport terminal. http://resolver.tudelft.nl/uuid:43cb57f4-c01d-45d1-8b62-94659abbc735

Fichter, K., Lüdeke-Freund, F., Schaltegger, S., & Schillebeeckx, S. J. D. (2023). Sustainability impact assessment of new ventures: An emerging field of research. *Journal of Cleaner Production, 384*, 135452. https://doi.org/10.1016/j.jclepro.2022.135452

Guzzo, D., Pigosso, D. C. A., Videira, N., & Mascarenhas, J. (2022). A system dynamics-based framework for examining circular economy transitions. *Journal of Cleaner Production, 333*, 129933. https://doi.org/10.1016/j.jclepro.2021.129933

Interaction Design Foundation – IxDF. (2015, December 6). *An Introduction to Action Research.* Interaction Design Foundation – IxDF. www.interaction-design.org/literature/article/an-introduction-to-action-research

Jones, P. (2014). Systemic design principles for complex social systems. In G. Metcalf (Ed.), *Social Systems and Design. Translational Systems Sciences* (Vol. 1). Springer. https://doi.org/10.1007/978-4-431-54478-4_4 https://doi.org/10.1007/978-4-431-54478-4_4

Jones, P. (2018). Contexts of co-creation: Designing with system stakeholders. In P. Jones & K. Kijima (Eds.), *Systemic Design. Translational Systems Sciences* (Vol. 8, pp. 3–52). https://doi.org/10.1007/978-4-431-55639-8_1

Kujala, J., Sachs, S., Leinonen, H., Heikkinen, A., & Laude, D. (2022). Stakeholder engagement: Past, present, and future. *Business & Society, 61*(5), 1136–1196. https://doi.org/10.1177/00076503211066595

Oberholzer, S., & Sachs, S. (2023). Engaging stakeholders in the circular economy: A systematic literature review. In J. Kujala, A. Heikkinen, & A. Blomberg (Eds.), *Stakeholder Engagement in a Sustainable*

Circular Economy (pp. 57–97). Springer International Publishing. https://doi.org/10.1007/978-3-031-31937-2_3

Özbay, İ., & Gokceviz, N. A. (2022). Towards zero-waste airports: A case study of Istanbul airport. *Journal of Material Cycles and Waste Management*, *24*(1), 134–142. https://doi.org/10.1007/s10163-021-01308-2

Pålsson, H., & Olsson, J. (2023). Current state and research directions for disposable versus reusable packaging: A systematic literature review of comparative studies. *Packaging Technology and Science*, *36*(6), 391–409. https://doi.org/10.1002/pts.2722

Parra, M. P. (2023). *Assessing the Status of Airport Collaborative Decision Making (A-CDM) Concept at European Airports*. Escola d'Enginyeria de Telecomunicació i Aeroespacial de Castelldefels, Universitat Politècnica de Catalunya. https://hdl.handle.net/2117/383721

Pizzi, S., Leopizzi, R., & Caputo, A. (2022). The enablers in the relationship between entrepreneurial ecosystems and the circular economy: The case of circularity.com. *Management of Environmental Quality: An International Journal*, *33*(1), 26–43. https://doi.org/10.1108/MEQ-01-2021-0011

Ritzer, G., & Jurgenson, N. (2010). Production, consumption, prosumption: The nature of capitalism in the age of the digital 'prosumer'. *Journal of Consumer Culture*, *10*(1), 13–36. https://doi.org/10.1177/1469540509354673

Sætra, H. S., & Selinger, E. (2024). Technological remedies for social problems: Defining and demarcating techno-fixes and techno-solutionism. *Science and Engineering Ethics*, *30*(6), 60. https://doi.org/10.1007/s11948-024-00524-x

Sassanelli, C., & Terzi, S. (2023). Circular economy and sustainable business performance management. *Sustainability*, *15*(11), 8619. https://doi.org/10.3390/su15118619

Schaar, D., & Sherry, L. (2010). Analysis of airport stakeholders. *2010 Integrated Communications, Navigation, and Surveillance Conference Proceedings*, J4-1–J4-17. https://doi.org/10.1109/ICNSURV.2010.5503233

Steg, L., & Vlek, C. (2009). Encouraging pro-environmental behaviour: An integrative review and research agenda. *Journal of Environmental Psychology*, *29*(3), 309–317. https://doi.org/10.1016/j.jenvp.2008.10.004

Tenhunen-Lunkka, A., Balatsas-Lekkas, A., Mouazan, E., Palola, S., Ngo, T., Salo, M., Hylkilä, E., Sundqvist, H., Luomala, H., Pennanen, K., Sorvari, K., Petänen, P., & Lahtinen, J. H. (2024). Implementing a circular business model for reusable packaging: Multidisciplinary learnings from reusable pizza packaging. *Sustainable Production and Consumption*, *48*, 62–83. https://doi.org/10.1016/j.spc.2024.05.006

Tjahjono, M., Tjahjono, B., Ünal, E., & Tran, T. H. (2024). Towards an improvement of airport retail waste management using the double diamond design process. In F. J. G. Silva, L. P. Ferreira, J. C. Sá, M. T. Pereira, & C. M. A. Pinto (Eds.), *Flexible Automation and Intelligent Manufacturing: Establishing Bridges for More Sustainable Manufacturing Systems* (pp. 946–953). Springer Nature. https://doi.org/10.1007/978-3-031-38165-2_109

van der Bijl-Brouwer, M., & Malcolm, B. (2020). Systemic design principles in social innovation: A study of expert practices and design rationales. *She Ji: The Journal of Design, Economics, and Innovation*, *6*(3), 386–407. https://doi.org/10.1016/j.sheji.2020.06.001

Van Der Tuin-Rademaker, A., Tschavgova, E., Van Maaren, C., Solis, S., Campisano, S., & Van Dam, S. (2024). Transforming waste management methods: A Dutch Airport's journey toward a circular economy through baseline measurements and strategic priority setting. *Frontiers in Sustainability*, *5*, 1356041. https://doi.org/10.3389/frsus.2024.1356041

Yin, R. K. (2014). *Case Study Research*. SAGE Publications.

Appendices A-C online material

4TU.ResearchData. Case study data: Reusable takeaway packaging system piloted at Norwegian airport. https://doi.org/10.4121/5921750a-aa77-461b-a4b1-090647626b1a

11 Digital Archiving with 3D Photogrammetry and On-Site Ideation for Enhanced Participation with Non-Humans

Tokushu Inamura, Melanie Sarantou, Kanako Tomisawa, and Oka Daisuke

Introduction

The shift towards human-animal engagement as a core goal of zoological gardens led to the introduction of petting zoos and other activities, enabling urban families to experience this type of interaction. An example from the literature is Tora Holmberg (2015, p. 1), who asks, 'What are the multispecies experiences and politics of living in a city?' From this perspective, understanding engagement is crucial as zoological gardens continually seek to improve their services and interactions. Engagement can be viewed as a 'critical phenomenon of interest' (de Vreede et al., 2019, p. 749). These authors define engagement across affective, behavioural, and cognitive domains, allowing stakeholders to target specific areas for improvement (de Vreede et al., 2019, p. 749).

This study focused on Guinea pigs at Omuta City Zoo (OCZ) in Omuta City, Japan. In this chapter, the authors intentionally capitalise the noun 'Guinea pig' following industry convention (Potter, 1984). The project explored sustainable enclosures, methods, and systems aimed at improving the Guinea pigs' quality of life. Beyond the socially oriented objectives of raising awareness among the Omuta community about local zoo activities and human-animal care relationships, the project specifically addressed future plans for enhanced animal husbandry facilities for over 60 Guinea pigs. These social animals live in groups and thrive through social interactions influenced by their environment and past experiences (Sachser, 1998). In the wild, they form herds and build deep social bonds, providing companionship and protection against predators (Asher et al., 2004). Such social activities are crucial aspects of their behaviour, observable at OCZ.

The study is based on the premise that engagement can be enhanced through participatory activities with zoo animals. It examined potential collaboration between humans, animals, AI, and digital tools to capture insights through mobile photography, fostering new forms of participation involving human, animal, and digital interactions.

The research investigates methods to improve engagement in zoos through participatory activities. It explores how collaboration between humans, animals, AI, and digital tools for capturing insights via mobile photography can promote new interactions. The key questions are: How can digital data sets of human-nonhuman collaboration help us understand, document, and enhance participation? What new forms of human-nonhuman participation can digital data generate?

The methodology employed was participatory arts-based research (PABR) (Nunn, 2022), integrating arts methods with values from postmodern, ethical, action-based, and futuristic research approaches for social inquiry (Finley, 2008, p. 71). Focusing on participation, Nunn

DOI: 10.4324/9781003667872-14

(2022, p. 251) describes PABR as a way to 'foster inclusive relationships.' This project adopted a posthuman, life-centred (Borthwick et al., 2022) approach to inclusion, moving away from predominantly human-centred methods. Supported by visual arts and design techniques like sketching and low-fidelity prototypes, a three-workshop process and observation generated digital visual data, which was partially developed into three-dimensional visual datasets.

The main findings were that the visual datasets resulted in imperfect 3D outcomes. Despite limitations, these data provided multidimensional documentation of concepts relevant to Guinea pig husbandry. The project's limitations include not having introduced the prototypes to the Guinea pigs themselves. However, the project continues until 2026, and future work will explore how the Guinea pigs respond to the prototypes.

Urban Spaces as Interfaces for Human-Nonhuman Creative Engagements

Animal husbandry is the study and practice of breeding, raising, and caring for animals. It is viewed from different perspectives, including agriculture, culture, economics, biogenetics, veterinary medicine, health, and the environment (Solona et al., 2023; Sinclair et al., 2018; van Asselt, 2019). Traditionally, animal husbandry literature has focused on production, yield, and cost-to-profit ratios. However, society now prioritises animal welfare and ethics more highly. The goal is to optimise farm animal production to meet the increasing global demand for food (Liu et al., 2023; Brasso et al., 2025). This involves considering agricultural, economic, biogenetic, veterinary, health, and environmental factors. Technological developments, such as the Internet of Things, introduce new dimensions to the field through tracking, health monitoring, automated feeding, and other applications (Rahman et al., 2024). Animal husbandry in zoological gardens encompasses more than just the animals themselves. Zoos are cultural institutions (Wolch & Emel, 1998) where animals are regarded as educators and ambassadors (Madden & Kearns, 2022). This promotes cooperation between people and animals, leading to new opportunities for collaboration, resilience, and biodiversity (Herrmann-Pillath et al., 2023). It can also enhance the visitor experience (Herrmann-Pillath et al., 2023). Such attitudes of care aim to balance the physical and mental well-being of both zoo animals and humans. Zookeepers are active participants in the social environment (Tuite et al., 2022). This new co-creative dynamic between species demonstrates how design can assist communities in collaborating with different species (Zhao & Zhang, 2023).

The focus is shifting from animal husbandry to collaborations and cohabitation involving multiple species, with innovative art and design approaches influencing this field (Liu, 2019). Art can transform interspecies relationships, fostering multispecies partnerships (Kozak, 2024). This partnership can extend to urban spaces, including communities and nonhuman and multispecies entities (Oliver, 2024). Art and design practices can incorporate animal agency in multispecies placemaking (Sachs Olsen, 2022). Art-based approaches can enrich ethical multispecies relationships through creative, participatory practices that prioritise animal agency and well-being (Krauth, 2022). This collaboration can extend to design that accommodates a more-than-human participatory environment with interspecies interaction, which can transform urban spaces.

The dynamics of interspecies social actions, such as human-nonhuman collaboration and communication, are essential for these cooperative practices in various settings and can be enhanced through multimodal approaches (Peltola & Simonen, 2024). Multispecies interactions foster shared experiences, deepen understanding of relationships, and highlight nonhuman perspectives that challenge an anthropocentric worldview (Dashper & Buchmann, 2020). Precision digital technologies, such as AI for health monitoring, automated feeding, and animal tracking, are now widespread. They facilitate real-time data collection, improving decision-making (Solona et al., 2023). Technological advancements add value and create opportunities for human-nonhuman collaboration via digital applications, engagement, and design (Miettinen & Sarantou, 2025). These include performance collaborations and visual communication, which can boost co-creation and multispecies placemaking.

Three-Dimensional Data Capturing in Human-Nonhuman

Three-dimensional (3D) capturing methods, such as photogrammetry, involve creating a 3D mesh by combining information from photographs. 3D capture techniques, including photogrammetry, have been tested in zoo enclosure design applications mainly to understand and gather data on existing spaces (Goh et al., 2017; Ritzler et al., 2025). However, these cases focus on understanding or digitalising existing spaces and enclosures. Regarding participatory engagement, opportunities exist before the actual spaces are completed. Prior work that focuses on earlier stages of capturing space shows that 3D photogrammetry for architectural models has been tested for architectural education and appears promising (Rashid et al., 2022). Abramov et al. have also compared different digital capturing techniques and applied them to engage people in the metaverse for participatory urban design (2024). Drones equipped with cameras have been used, alongside the technique of Gaussian splatting. Rather than representing spaces with a 3D mesh, this method uses data from image orientations derived from photogrammetry but differs in how it handles geometry and visualisation (Abramov et al., 2024). In humanitarian applications, these techniques have been validated, particularly for engaging methods of post-disaster urban mapping by citizen scientists (Pezzica et al., 2019). Such mapping facilitates participatory engagement in imagining future designs of places in transition. Although there is prior work within zoo contexts and human participatory projects, research exploring the potential of 3D capturing technologies like photogrammetry in participatory human-nonhuman collaboration remains limited.

Methodology

Using an ABR methodology (Leavy, 2017), the project gathered extensive visual data that was analysed to understand the OCZ's needs for developing new enclosures and interactive spaces for engagement with the Guinea pigs and the future services it aims to offer its audiences. The methods, data, and documentation tools will be discussed hereafter.

Arts-based methods (ABMs): ABMs, such as sketching and low-fidelity prototype creation, were employed in a three-workshop series, supported by observation, note-taking, and photographic documentation. The arts-based technique used was sketching; however, low-fidelity prototyping was also implemented in two of the workshops. Storytelling was used to explore the ideas and visions of the participants. Participants' stories were recorded using a digital mobile phone sound recorder and video. They used sketching paper, cardboard boxes, and various craft materials, including sticky tape, glue, paints, Posca pen markers, scissors, and stickers. In the second workshop, many children interacted with the balloons displayed in the venue. The balloons were features of the work environment provided by the Picture Book Museum, officially known as the *Oomutashi Tomodachiya Ehonbijutukan* (OTE). The museum is located within the OCZ grounds where the workshop took place. The sketches and prototypes were digitally documented using mobile phone cameras. The prototypes from the second workshop were recorded using a mobile phone and the quick photo-capturing Polycam 3D app. This app required at least 100 photos to generate three-dimensional models of the prototypes (see Figure 11.4).

Contemplative practices: Complementary to ABMs were generative contemplative practices, specifically guided visualisation through meditation and mind-wandering, as forms of creative thinking (Davidson & Dahl, 2016; Duerr & Bergman, n.d.; Fox & Beaty, 2019). Contemplative practices are practical methods that aim to create prolonged states of creativity or well-being (Davidson & Dahl, 2016). Meditation was utilised as a generative contemplative practice across all three workshops to stimulate visualisations of the Guinea pigs and potential future living scenarios they might face. The meditation activities lasted between three and four minutes.

Observations: The participants and researchers engaged in observations, such as animal husbandry checks, visiting the Guinea pig enclosure and its activity area at the zoo, and participating in a programme to observe Guinea pigs and their surroundings. Participants used sticky notes and illustrations to record their thoughts and creative ideas for the future Guinea pig enclosure. The researchers observed and documented the activities using digital photos and videos. On-site digital documentation created detailed visual datasets, capturing insights and potential solutions. Observations were recorded using sticky notes, sketching, and a digital whiteboard tool on the Miro platform.

The Workshops and Participants: We organised three workshops at OCZ and the adjacent Picture Book Museum, OTE. The project involved zoo staff, keepers, children aged three to nine years, and adult zoo visitors.

- Workshop A included 16 OCZ/OTE staff members to clarify the intentions and visions of the zoo stakeholders from a husbandry perspective, focusing on future Guinea pig enclosures. The initial two-hour session also aimed to build support for the project. Conducted in March 2024.
- Workshop B lasted 1 hour and 30 minutes, with 12 children aged three to nine years and their parents or carers. Its goal was to foster a connection between children and the Guinea pigs and to stimulate their imagination regarding potential enclosure designs. Conducted in November 2024.
- Workshop C, lasting 1 hour and 45 minutes, had similar aims to Workshop B but focused on xx (add amount) adult visitors. Eight adults and one child participated. Invitations were issued publicly, and further invitations were extended via OCZ/OTE/University local networks, including picture book reading volunteers, regular zoo workshop attendees, and local preschools. Conducted in December 2024.

The content of the workshops is summarised in Table 11.1. The workshops were developed through ongoing reflection. Building on the first workshop with staff, an overarching narrative of Guinea Pig Town was created as a concept to frame the workshops. Each workshop focused on different levels of human participants and addressed the development of enclosures from both husbandry and visitor engagement perspectives.

The three workshops provided different data sets from the three stakeholder groups, offering various perspectives that the zoological garden can use to develop new engagement programmes and enhance facilities.

Table 11.1 The three-workshop approach

	Workshop A	*Workshop B*	*Workshop C*
Participants	OCZ staff and animal caretakers	Children aged 3–9 years	Mainly Adults and one child zoo visitors
Workshop process	Step 1: Team forming Step 2: Need/opportunity finding Step 3: Visualisation and developing ideas Step 4: Presentations and reflections	Step 1: Observation of animal behaviour Step 2: Embody and imagine. Step 3: Making Guinea Pig Town Step 4: Presentation	Step 1: Observation Step 2: Imagining and Sketching Step 3: Making Guinea Pig Town Step 4: Presentation
Methods	Meditative guided visualisation: Sketching, note-taking, storytelling	Meditative guided visualisation; performative and Embodied becoming ('being a Guinea pig'), Low-fidelity three-dimensional prototype-making	Sketching, note-taking, low-fidelity three-dimensional prototype-making, storytelling
Data Outcomes	Sketches and notes, documentary photographs	Low-fi prototypes, 9 sets of 3D data sets, documentary photographs	Sketches, notes, low-fi prototypes, documentary photographs, 8 3D data sets

Data Documentation and Analysis: For content analysis, the research team strived to understand the terms used for note-taking to establish meanings and relationships of certain words, themes, or concepts (Krippendorff, 2004). On-site digital documentation in Workshops B and C was employed to capture prototypes of the children and adults, respectively. The prototypes were immediately documented using a mobile phone using Polycam and by taking multiple images of the prototype while rotating the camera around the object. The images were processed using the 3D photogrammetry and Gaussian Splat modes. Based on the 3D prototypes, a set of 16 three-dimensional data sets was captured, providing new solutions for participatory documentation and generating new opportunities for archiving insights and potential solutions for new animal enclosure design at OCZ. These activities resulted in creative and collaborative outputs that involved all age groups.

Using the four-step process of Seymour (2023, para. 7), visual analysis was employed to analyse the collected visual and notated data, establishing the meanings and relationships of the participants' symbols, themes, or concepts. The visual data was analysed using the four steps: *Identify* (how art or design elements are used), *Describe* (how elements in the composition function in their entirety), *Connect* (how the expression is connected to its broader cultural, historic, and social networks) and *Interpret* (what the expression communicates and the reaction to its interpretation). From the four-step analysis, Table 11.2 was developed.

The findings will illustrate that the documentation methods delivered imperfect digital visual data (Figure 11.4). However, these outcomes delivered valuable insights into the visions for engaging with the animals in zoological gardens. The data, although imperfect, were developed into three-dimensional visual data sets that illustrated the visions and hopes of three groups of participants for enhanced engagement with the OCZ and its Guinea pig population.

Table 11.2 Results of the visual analysis using the 16 visual data sets documented through 3D data capture

	Description highlights	*Key features*	*Theme / inspiration*
1	Circus flags, slide, stairs, artwork on the walls.	Flags, stairs, slide, artwork, rocks, trees, flagpoles, floral decorations Circus and/or market with decorative art	Circus and/or marketplace featuring decorative art
2	Arched gazebo, high arches, resting spaces	Arch-shaped gazebo shelter, elevated pointed roof, resting place	Shelter-focused, aesthetic arches
3	Love heart décor, door with a handle, windows, table	Heart motifs, window/door access, small furniture	Whimsical caring/love theme
4	Two-box dwelling, double door, soft materials inside and on top	Entryways, soft surfaces, stools on top and inside for comfort	Functional double-box design for comfort
5	White balloon over green entrance, colourful flowers	Balloon decorated entrance, interior grassy green area, bright paper flowers on walls	Playful garden-inspired with balloons
6	Double-story enclosure, open bottom, slide from the top floor at the end of a staircase	Shady open bottom, enclosed top hideaway, slide, decorative elements, block-style play structures	Pagoda-inspired play structure
7	Cutaway double-story enclosure, a mix of shaded and open spaces, furniture and block shales scattered around	Red balloon outside, open sides, blocks, partial enclosures for airflow	Indoor-outdoor blend with playful red accents
8	Two white soft areas for resting (may be beds), a high wall-mounted box-shaped hideaway, purple and pink décor	Window openings, hideaway, colourful wall elements	Pastel and soft colour palette with decorative purpose
9	Obstacle course in the shape of an enigma, flower, shading	High and low elevations as walk/run paths, shady areas, flower features, box supports	Maze-like playful design
10	Black and white cylindrical tunnel-like protrusions	Cylindrical shapes, multiple entry tunnels, monochrome colour palette	Geometric and modern
11	River, two bridges, triangular green area, domed roof with low ends for access to the top of the roof	Dome structure, bridges, sloped roof, sheltered areas, hiding areas	Nature and landscape mimicry
12	Green grassy area, mounds, trees, yellow food-rich pastures	Hills, trees, hidden walk--through spaces, water elements, circular food and flower patches	Natural ecosystem, hills and fields theme
13	Colourful arch tunnel entrance, fan-like blue object (may be a satellite dish or some mechanical addition like a wind turbine or a fan)	Arch entry, rounded blue top, garden with red/green colour palette	Garden-technology blend
14	Dome-like Guinea pig-shaped dwelling with whiskers, black and white spots to mimic the Guinea pigs from the zoo	The entire structure mimics a Guinea pig, with whisker decoration, front snoot-like opening	Literal Guinea pig-shaped enclosure

(*Continued*)

Table 11.2 (Continued)

	Description highlights	*Key features*	*Theme / inspiration*
15	Complex multi-structure layout, grassy garden area, tower-like element, possibly a flagpole	Tall multi-entry tower, roof-top lookout, green surroundings	Vertical exploration and lookout space
16	Square shape with 4 dome-shaped entrances, a blue roof-top canopy, internal garden	Blue pagoda-style roof, grassy floor, window/door access, second level with walk-through openings	Balanced indoor-outdoor with ventilation

Results

The three workshops produced extensive visual datasets, as shown in Figures 11.1–11.4, capturing participants' behaviours, expressions, hopes, concerns, and visions regarding their perceptions of future enclosures for the Guinea pig population at OCZ. For example, from the visual analysis (Table 11.2), the following themes were identified:

Interactive features: Obstacle courses, tunnels, slides, and bridges indicate an emphasis on activity and stimulation for the Guinea pigs.

Playfulness and care: The use of playful elements and symbols of care, such as slides, tunnels for crawling or sliding, love hearts, and areas for shelter, rest, and feeding.

Natural elements: Many incorporated green spaces, mounds, trees, and even rivers to emulate a more natural habitat.

Use of colour and decoration: Participants often utilised bright and primary colours, along with flowers and balloons, to decorate and enhance visual appeal.

Multi-level designs: Several enclosures feature double-story configurations or elevated play and hide areas, highlighting a focus on vertical exploration.

Use of unique shapes: Some designs directly mimic animals (e.g., Guinea pig-shaped) or employ abstract forms like maze-like structures or domes for creative themes.

Alongside the photo data sets capturing the three workshop processes and illustrations, low-fidelity prototypes and participant notes were used to document the participants' experiences. Using rapid-collection three-dimensional photo sets, ABMs such as sketches, drawings, and photographic documentation allowed for immediate on-site ideation during the three-workshop series. These findings are shown in Figures 11.1–11.4. As seen in Figure 11.2, observation of the animals was included in the last two workshops, which targeted non-zoo professionals. These exercises were followed by contemplative practices, including meditative and role-playing activities, to encourage embodied and empathetic states in the participants. Participants were also asked to learn the name of a single Guinea pig from zoo staff. This prompting led to making and storytelling activities, reflected in the imaginative prototypes created during the workshop.

The bottom two images from Figure 11.4 depict two angles of a prototype developed in Workshop B. Although all data capture was processed using 3D photogrammetry, one prototype was tested for comparison purposes using Gaussian splatting (Figure 11.4 bottom left) and photogrammetry (Figure 11.4 bottom right). The features were consistent; however, Gaussian splatting preserved more of the surrounding workshop space, colours, and noisy textures. The artefact highlights key aspects of empathy and embodiment. For instance, it includes a slide and stairs, illustrating the playful imagination of the child participant and their consideration—based on observation—for how Guinea pigs might move playfully and practically within the space. These elements were central to the researchers' thematic analysis, focusing on themes such as empathic imagination and shared play between children and Guinea pigs. These themes emerged from identifying features that reflected the children's personal experiences of play and enjoyment, such as stairs, slides, maze-like structures, floral decorations, and green areas (see Table 11.2). Describing these elements and their functions helped explore and link these themes to future design goals and aspirations for the upcoming Guinea pig enclosure at OCZ. Furthermore, these elements could be related to activities—for example, using the noun "slide" and connecting it to participatory play—and interpreted by OCZ staff as they design new animal enclosures in the future.

Discussion

ABMs encourage the development of new ideas and visions through spontaneous creative expression. These methods support a lively process where possibilities may remain unknown until they emerge, shaped by the unique environment. When combined with contemplative practices, ABMs can stimulate creative and imaginative thinking (Dobson & Christoff, 2020). ABMs also provide opportunities for healing and play (Malchiodi, 2013). They offer tactile and embodied interactions that invite fresh perspectives and connect zoo professionals, animals, and visitors in a cycle of co-creative development. Such approaches open pathways for reimagining the future and inspiring innovation (Eaves, 2014).

Since the materials used were suitable for enabling ABMs and low-fidelity prototyping, they were not chosen to be Guinea pig-proof. The prototypes were not tested by presenting them to Guinea pigs to observe their reactions, because the materials would likely be unsuitable for Guinea pig enclosures. Each model is available for analysis from a husbandry perspective by professional zoo staff. Therefore, the absence of a documented evaluation of the prototypes should not be seen as a failure at this stage of the Guinea Pig Town ideation process. However, creative archiving of the development of low-fidelity prototypes could facilitate the creation of Guinea pig-friendly spaces in the future. These development processes could actively explore new avenues for nonhuman engagement in decision-making about their living environments.

All designs are imperfect and in the rough prototyping stage. The tangible outcomes of ABMs are examples of the zoo and the Guinea pigs' enclosure's future. Participants' ideas are vital for engaging with the well-being of both human and nonhuman stakeholders. The digital data generated during the project can inspire further discussion based on identified values, features, or design characteristics. However, imperfect data can also spark future inspiration and lead to miscommunication. Torrisi and Hall (2013) note that miscommunication can foster innovation. Their earlier research on group design work explains how the C-sketch technique allows participants to 'freely reinterpret each other's early ideas by exploiting miscommunication to generate novel design concepts'. Similarly, imperfect 3D photogrammetry and Gaussian splatting, a three-dimensional representation of real-world data captured through a volume rendering method, can harness the imaginations of engaged beings. The ability to share and browse 3D data permits the preservation of spatially explorable datasets. In Figure 11.4, bottom left, we see the prototype rendered with Gaussian splatting, which displays good visual quality. Conversely, the bottom right image shows the same model, but with a form resembling deconstructive architecture. Both methods demonstrate the potential of creative and participatory archiving. The study finds that combining participatory ABMs with digital tools like photogrammetry can:

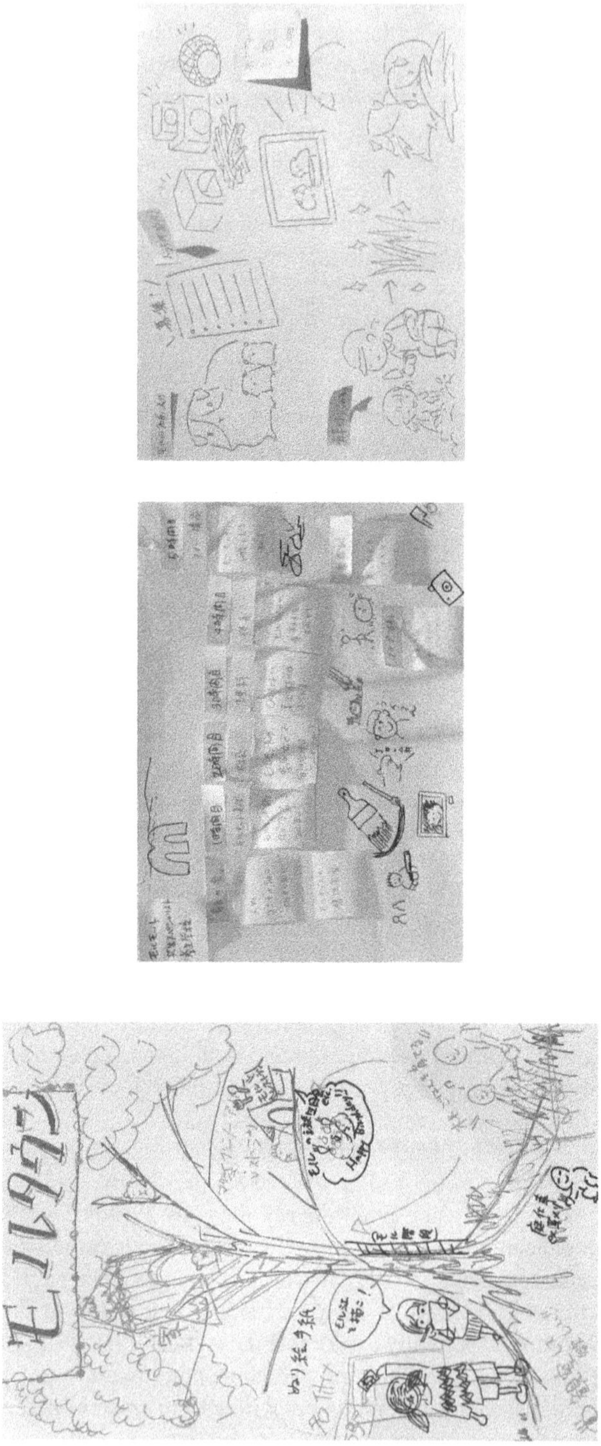

Figure 11.1 Workshop A with 17 OCZ/OTZ staff members and animal caretakers. Photograph by the author.

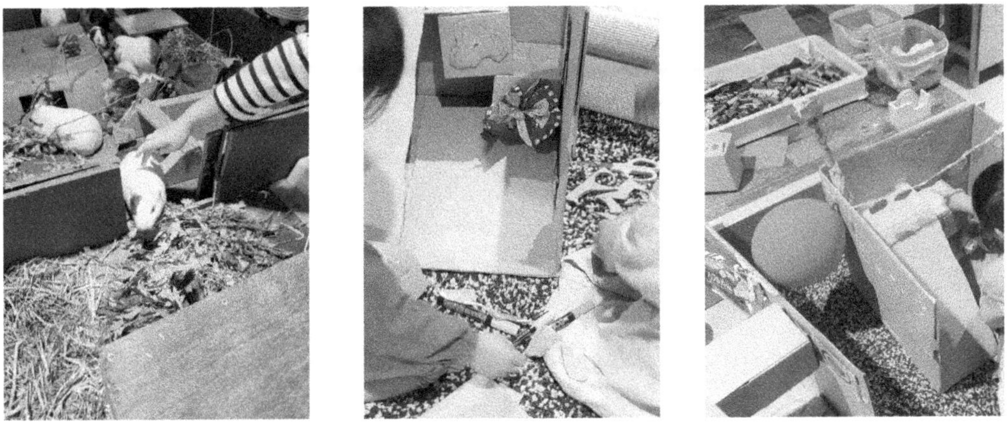

Figure 11.2 Workshops B with children ages 3–9, assisted by parents. Photograph by the author.

Figure 11.3 Workshop C with adult participants. Photograph by the author.

- Enhance ethical and inclusive animal care practices by integrating ABMs with generative contemplative practices to promote engagement.
- Encourage meaningful human–animal–digital collaborations through digital tools, rapid photo capture, and other documentation methods to create multimodal data for creative interpretation.
- Offer educational experiences for all ages, enriched by animal-human behaviour observation and guided creative techniques that stimulate innovative thinking and foster creative expression for sustainability.
- Highlight the significance of visual and material engagement in zoo environments for co-designing future enclosures.

The aim of this project was twofold: firstly, to facilitate long-term archiving, and secondly, to promote future cycles of prototyping and participation between nonhuman and human entities. Regarding the value of 3D data capture in archiving, considerations are relevant for participatory rapid prototyping and quick result documentation. For example, archiving can address the fragility of prototypes, which arises from their transient nature and storage limitations. Prototypes are often discarded after a short period, as they tend to be produced through quick, makeshift methods. Therefore, the durability of prototypes is questionable, and concerns exist regarding the feasibility and cost of long-term storage. 3D capturing techniques have been demonstrated

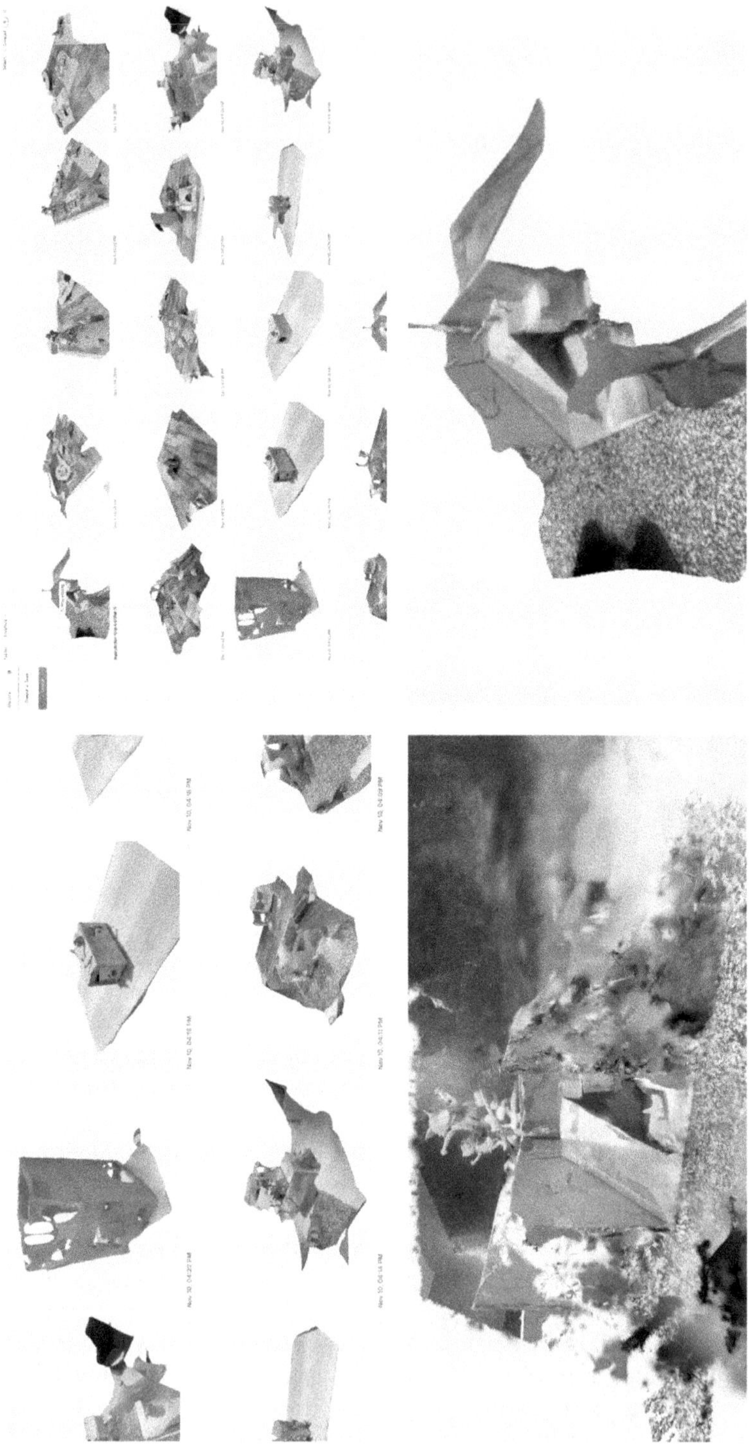

Figure 11.4 One set of the fast photo capture was produced using the Polycam 3D application. Photograph by the author.

to compensate for the loss of materiality by preserving participants' memories and recording their ideas for archiving. This helps reduce design waste.

Conclusion and Future Work

Through this study, we have introduced new forms of engagement that foster human–animal–digital collaborations via participatory and contemplative ABMs. We examined how the low-fidelity prototypes from participatory workshops, captured by rapid mobile photogrammetry and Gaussian splatting, despite producing imperfect data, are valuable for inspiring new dynamics and approaches to interspecies participation. These imperfections serve as a form of visual archival documentation rich in interpretive potential. Such archives not only record the co-creative process but also support future iterations of prototypes that envisage safe, stimulating enclosures—reimagined urban spaces—for interspecies engagement between Guinea pigs and humans. Moving forward, immediate efforts are directed towards the co-development and co-evaluation of these enclosures through husbandry and ABR perspectives. More broadly, we hope that these situated methods of care can inspire new participatory approaches for broader groups of animals within zoological and urban ecological contexts.

References

Abramov, N., Lankegowda, H., Liu, S., Barazzetti, L., Beltracchi, C., & Ruttico, P. (2024). Implementing Immersive Worlds for Metaverse-Based Participatory Design through Photogrammetry and Blockchain. *ISPRS International Journal of Geo-Information, 13*(6), 211. https://doi.org/10.3390/ijgi13060211

Asher, M., de Oliveira, E., & Sachser, N. (2004). Social System and Spatial Organisation of Wild Guinea Pigs (Cavia Aperea) in a Natural Population. *Journal of Mammalogy, 85*(4), 788–796. https://doi.org/10.1644/BNS-012

Brassó, L. D., Komlósi, I., & Várszegi, Z. (2025). Modern Technologies for Improving Broiler Production and Welfare: A Review. *Animals, 15*(4), 493.

Dashper, K., & Buchmann, A. (2020). Multispecies Event Experiences: Introducing More-Than-Human Perspectives to Event Studies. *Journal of Policy Research in Tourism, Leisure and Events, 12*(3), 293–309.

Davidson, R. J., & Dahl, C. J. (2016). Varieties of Contemplative Practice. *JAMA Psychiatry, 74*(2), 121–123. https://doi.org/10.1001/jamapsychiatry.2016.3469

de Vreede, T., Andel, S. A., de Vreede, G.-J., Spector, P., Singh, V., & Padmanabhan, B. (2019). What Is Engagement and How Do We Measure It? Toward a Domain Independent Definition and Scale. In *Proceedings of the 52nd Hawaii International Conference on System Sciences*, p. 749. https://hdl.handle.net/10125/59515

Dobson, C., & Christoff, K. (2020). Productive Mind Wandering in Design Practice. In *Creativity and the Wandering Mind*, pp. 271–281. Academic Press. https://doi.org/10.1016/B978-0-12-816400-6.00012-2

Duerr, M., & Bergman, C. (n.d.). *What Is Contemplative Practice?* Contemplative Resource Centre, University of Colorado, Accessed April 3, 2025, from: www.colorado.edu/center/contemplativeresource/what-contemplative-practice

Eaves, S. (2014). From Art for Arts Sake to Art as Means of Knowing: A Rationale for Advancing Arts-Based Methods in Research, Practice and Pedagogy. *Electronic Journal of Business Research Methods, 12*(2), 154–167.

Finley, S. (2008). Arts-Based Research. In Knowles, J. G., Cole, A. L. & Finley, S. (Eds.), *Handbook of the Arts in Qualitative Research: Perspectives, Methodologies, Examples, and Issues*, pp. 71–81.

Fox, K. C., & Beaty, R. E. (2019). Mind-Wandering as Creative Thinking: Neural, Psychological, and Theoretical Considerations. *Current Opinion in Behavioral Sciences, 27*, 123–130. https://doi.org/10.1016/j.cobeha.2018.10.009

Goh, C., Blanchard, M. L., Bates, K. T., Manning, P. L., & Crompton, R. H. (2017). A Novel Approach to Studying Enclosure and Support Usage in Siamangs: Using a 3D Computer Model. *Journal of Zoo and Aquarium Research*, *5*(3), 109–115.

Herrmann-Pillath, C., Sarkki, S., Maran, T., Soini, K., & Hiedanpää, J. (2023). Nature-Based Solutions as More-than-Human Art: Co-Evolutionary and Co-Creative Design Approaches. *Nature-Based Solutions*, *4*, 100081.

Holmberg, T. (2015). *Urban Animals: Crowding in Zoocitie* (1st ed., p. 1). Routledge. https://doi.org/10.4324/9781315735726

Kozak, L. (2024). Fostering an Inclusive Multispecies Social Pedagogy: Rethinking Ethical Human-Animal Relations in Art. *International Journal of Social Pedagogy*, *13*(1), 1–12.

Krauth, A. (2022). *More-Than-Human Creative Practice: Approaches to Making Interactive and Digital Art as Enrichment for Wild Flying Foxes and Domesticated Dogs*. (Doctoral dissertation, Queensland University of Technology).

Krippendorff, K. (2004). *Content Analysis: An Introduction to Its Methodology*. Sage.

Leavy, P. (2017). *Research Design: Quantitative, Qualitative, MixedMmethods, Arts-based, and Community-Based Participatory Research Approaches*. Guilford Press.

Liu, G., Qin, P., Cheng, X., Wu, L., Wang, R., & Gao, W. (2023). Ursolic Acid: Biological Functions and Application in Animal Husbandry. *Frontiers in Veterinary Science*, *10*, Article 1251248. https://doi.org/10.3389/fvets.2023.1251248

Liu, S. Y. (2019, June). Designing With, Through, and for Human-Nature Interaction. In *Companion Publication of the 2019 on Designing Interactive Systems Conference 2019 Companion*, pp. 101–104. ACM.

Madden, L., & Kearns, R. (2022). Performing Encounters (and EncounteringP) at Auckland Zoo. *New Zealand Geographer*, *78*(2), 123–133. https//doi.org/10.1111/nzg.12338

Malchiodi, C. A. (2013). *Handbook of Art Therapy* (2nd ed). The Guilford Press.

Miettinen, S., & Sarantou, M. (2025). Visual Communication through Performance Collaborations. *Frontiers in Communication*, *10*, 1–14. https://doi.org/10.3389/fcomm.2025.1476289

Nunn, C. (2022). The Participatory Arts-Based Research Project as an Exceptional Sphere of Belonging. *Qualitative Research*, *22*(2), 251–268.

Oliver, C. (2024). Dancing with Foxes: More-than-Human Design in 'The Chicken City'. *Transactions of the Institute of British Geographers*, *50*(2), e12695. https://doi.org/10.1111/tran.12695

Peltola, R., & Simonen, M. (2024). Towards Interspecies Pragmatics: Language Use and Embodied Interaction in Human-Animal Activities, Encounters, and Narratives. *Journal of Pragmatics*, *220*, 15–19.

Pezzica, C., Piemonte, A., Bleil de Souza, C., & Cutini, V. (2019). Photogrammetry as a Participatory Recovery Tool after Disasters: A Grounded Framework for Future Guidelines. *International Archives of the Photogrammetry, Remote Sensing and Spatial Information Sciences*, *XLII-2/W15*, 921–928. https://doi.org/10.5194/isprs-archives-XLII-2-W15-921-2019

Potter, E. F. (1984). On Capitalization of Vernacular Names of Species. *The Auk, 101*(4), 29.

Rahman, A., Hussain, S. A., & Baig Z, T. (2024). Internet of Things (IoT) in Animal Healthcare and Location Tracking. In Godihal, J. H. (Ed.), *Connected Horizons Exploring IoT Applications in Infrastructure*. Environment and Design, IIP Series, pp. 1–9. https://doi.org/10.58532/nbennurcech1

Rashid, A. F. M., Husin, R., & Noor, M. P. M. (2022). Preservation of Architectural Model Into 3 Dimensional Digital Form with the Methods of Photogrammetry. *International Journal of Infrastructure Research and Management*, *10*(1), 72–77.

Ritzler, C. P., Mather, S. V., Johnson, I. T., Dennis, P. M., Lukas, K. E., & Koester, D. C. (2025). Using 3D Photogrammetry to Quantify Usable Space in Zoo-Housed Primate Habitats. *American Journal of Primatology*, *87*(2), e70002.

Sachs Olsen, C. (2022). Co-Creation Beyond Humans: The Arts of Multispecies Placemaking. *Urban Planning*, *7*(3), 315–325. https://doi.org/10.17645/up.v7i3.5288

Sachser, N. (1998). Of Domestic and Wild Guinea Pigs: Studies in Sociophysiology, Domestication, and Social Evolution. *Naturwissenschaften*, *85*(7), 307–317. https://doi.org/10.1007/S001140050507

Seymour, V. (2023). How to Look at Art and Understand What You See. JSTOR Daily. Accessed 2025 from: https://daily.jstor.org/how-to-look-at-art-and-understand-what-you-see/

Sinclair, M., Derkley, T., Fryer, C., & Phillips, C. J. (2018). Australian Public Opinions Regarding the Live Export Trade Before and After an Animal Welfare Media Exposé. *Animals*, *8*(7), 106.

Torrisi, V. S., & Hall, A. (2013). Missing Miscommunications in Interdisciplinary Design Practice. In *DS 76: Proceedings of E&PDE 2013, the 15th International Conference on Engineering and Product Design Education*, Dublin, Ireland, 05-06.09. 2013. pp. 581–586.

Tuite, E. K., Moss, S. A., Phillips, C. J., & Ward, S. J. (2022). Why Are Enrichment Practices in Zoos Difficult to Implement Effectively? *Animals*, *12*(5), 554. https://doi.org/10.3390/ani12050554

van Asselt, M. (2019). *Conflicts between Chicken Welfare and Public Health Risks in Poultry Husbandry: Stakeholders' Perceptions and Moral Convictions* (Doctoral dissertation, Wageningen University and Research).

Wolch, J., & Emel, J. (Eds.). (1998). *Animal Geographies: Place, Politics and Identity in the Nature-Culture Borderlands*. Verso.

Zhao, X., & Zhang, D. (2023). Design Activating Agency: A Study on Rural Community Co-creation in China Under Non-Anthropocentrism. In De Sainz Molestina, D., Galluzzo, L., Rizzo, F., Spallazzo, D. (Eds.), *IASDR 2023: Life-Changing Design* (pp. 1–14), 9–13 October, Milan, Italy. https://doi.org/10.21606/iasdr.2023.447

Section IV

Future Smart Cities and Transportation

12 Designing Mobility Futures through Strategic Citizen Engagement

Manuela Quaresma

Introduction

The rapid urbanization of major cities in the Global South (UNDESA, 2018) presents a series of complex challenges related to transportation, infrastructure, and social equity. In Rio de Janeiro, Brazil, these issues are especially pressing: insufficient public services, lack of integration between transport systems, safety concerns, and the marginalization of peripheral communities all contribute to a fragmented and often exclusionary urban mobility experience. As the promise of "smart cities" gains traction globally, the need to ground innovation in the lived realities and aspirations of citizens becomes increasingly urgent.

Urban mobility, as a central component of urban life, not only reflects infrastructural limitations but also raises deeper questions of inclusion, sustainability, and governance. The traditional top-down approaches to mobility planning have failed to address many of these systemic issues, especially in cities with deep social and spatial inequalities. In response, participatory and citizen-centered design methods have emerged as essential tools for rethinking how we engage with urban complexity, and some studies have already demonstrated the effectiveness of citizen participation in the development of products and services for smart cities (Noring et al., 2018; Khan et al., 2017; Kudo, 2016; Gebauer et al., 2010).

This chapter investigates how participatory design methods, combined with future-oriented approaches, can support the development of innovative public services and policies in urban mobility. The main objective of the research was to identify current and future citizen needs and desires and, based on that, to co-develop actionable solutions, including service innovations, product-service systems, and policy proposals, that respond to those challenges. By doing so, the study sought to anticipate structural problems in urban mobility from the citizen's perspective and offer concrete pathways to address them in a sustainable and inclusive manner. Two guiding research questions informed this process: (1) What are the primary needs and desires of citizens regarding urban mobility today and in the future? and (2) What types of systems and services could be envisioned from the perspective of these citizens to meet those expectations? The rationale behind this research stems from the growing urban population and the need to reduce car dependency, carbon emissions, and inequality in access to transportation. In this context, involving citizens in co-design processes became essential for imagining more equitable and adaptive solutions.

To address these questions, the study employed a dual-stage participatory process that integrates speculative and strategic design approaches. The first stage focused on envisioning desirable futures through speculative workshops; the second applied a collaborative backcasting approach to identify actionable pathways toward those futures. By placing citizens at the center of both visionary and strategic thinking, the methodology repositions engagement not as passive

DOI: 10.4324/9781003667872-16

consultation, but as an active, creative, and structured process for shaping public services and policies.

This approach contributes to the broader discussion on how design can support democratic innovation and citizen-driven transformation in urban contexts. By combining co-design and foresight methods, it offers a novel strategy for engaging people in the planning of mobility systems grounded in real human needs.

Smart Mobility, Urban Futures, and the Role of Citizen Participation

The concept of smart cities emerged as a response to the growing complexity of urban life, particularly under pressures such as rapid urbanization, environmental degradation, and the expansion of metropolitan populations. Early definitions emphasized the application of information and communication technologies (ICTs) to optimize infrastructure, improve service efficiency, and enhance resource management (European Commission, 2020). In this framing, "smartness" was often associated with automation, connectivity, and data-driven control systems applied to urban sectors such as transportation, energy, and security.

Over the past decade, however, this notion has evolved. Authors such as Kirimtat et al. (2020) and institutions like the OECD (2020) have contributed to a more nuanced view, shifting the emphasis from technical performance to social responsiveness. A truly smart city, in this expanded understanding, integrates technology with inclusive governance, environmental responsibility, and citizen well-being. This transition reflects a broader shift from a techno-centric paradigm to one that values transparency, civic participation, and equitable access to urban opportunities (Morozov & Bria, 2019).

Within this evolving framework, smart mobility has also undergone a conceptual expansion. Initially centered on technological innovation, like autonomous vehicles, integrated ticketing, real-time monitoring, and intelligent infrastructure, smart mobility is now increasingly framed as a human-centered, multimodal system. Vallet et al. (2020) argue that mobility must be seen not only as transportation technology but as a social and experiential phenomenon, shaped by everyday routines, subjective needs, and social roles.

This reframing positions mobility as a right, essential for accessing education, health, employment, and participation in urban life. Particularly in Latin American cities, where mobility systems often reflect spatial segregation and infrastructural neglect, a smart mobility strategy must be attentive to systemic inequalities and everyday vulnerabilities. Walking, cycling, riding public transit, or combining different modes are not merely logistical choices but expressions of spatial justice and inclusion.

In this context, there is the concept of the "15-minute city," which is a city where citizens can carry out their daily activities within 15 minutes of their homes, whether going to work, shopping, attending school, or engaging in any other daily activity (Moreno, 2020). Walking is favored in this concept, while micromobility and high-speed public transportation support longer trips. This model challenges car-dependent infrastructures and supports more localized, human-scale, and sustainable patterns of mobility.

Micromobility encompasses bicycles, scooters, and other compact vehicles and plays an important role in diversifying urban transport options. According to the National Association of City Transportation Officials (NACTO, 2018), it offers affordable, flexible, and low-emission alternatives that can complement public transit and close critical gaps in the mobility network. However, its successful adoption depends on safe, continuous, and connected infrastructure, especially in cities like Rio de Janeiro, where topography, security, and climate pose significant design challenges.

Shared mobility further expands the possibilities of inclusive urban movement. Encompassing bike-sharing, car-sharing, ride-hailing, and other non-ownership models, shared mobility has been identified by Shaheen et al. (2016) as a promising strategy to reduce emissions and dependence on private vehicles. It also has spatial implications: fewer private cars allow for the reallocation of urban space to green areas, public facilities, or pedestrian zones. Still, as in other mobility innovations, success depends on integration across systems, equitable access, and trust, especially in contexts marked by insecurity and social fragmentation.

Despite the growing discourse around innovation and inclusion, structural barriers persist in many Latin American cities. As Borba et al. (2022) note, Rio de Janeiro's transportation system is fragmented, inefficient, and unequal, with multiple poorly integrated modes of buses, metro, trains, ferries, and BRTs. Peripheral areas remain underserved, while mobility options are often insecure and stressful, especially for women and youth communities. Rio's challenging geography further exacerbates modal fragmentation, and violence compounds the emotional and physical strain of urban movement.

In this scenario, mobility reflects broader patterns of inequality, determining not just how people move, but who has access to opportunities and full urban citizenship. Addressing such complexity demands more than infrastructure; it requires a paradigm shift toward participatory, situated, and socially attuned strategies.

Faced with the limitations of centralized and technocratic planning, there is an urgent need to involve citizens as co-creators of urban mobility systems. This is not only a methodological choice but also a democratic imperative. Co-design, as defined by Sanders and Stappers (2008), offers a collaborative framework in which designers and non-designers jointly develop solutions. It recognizes that people are experts in their own experiences and that their situated, often tacit knowledge is essential to creating meaningful systems.

Co-design shifts the focus from isolated problem-solving to collective transformation (Mattelmäki et al., 2011). It bridges gaps between lived realities and institutional strategies, especially in cities where state interventions have been fragmented or inadequate. In this process, citizens move from passive informants to active participants, articulating not just needs but also values and visions for the future.

The potential of co-design expands further when combined with futures thinking. Rooted in futures studies (Gidley, 2017), these approaches invite participants to speculate on what could or should happen, moving beyond short-term fixes to long-term transformation. They offer tools for collective imagination and anticipation, allowing communities to explore scenarios, opportunities, and risks that are not yet visible in the present.

Vallet et al. (2020) underscore that futures thinking is especially powerful in complex domains like urban mobility, where conflicting interests, interdependencies, and uncertainty prevail. These approaches allow people to articulate aspirations not only in terms of functionality but also in cultural, emotional, and ethical terms. The goal is not to predict but to expand the space of possibility.

Among the most relevant futures tools is the method of backcasting—a strategic approach that begins with a normative vision of a desirable future and works backward to identify the structural changes and actions needed to reach it (Quist & Vergragt, 2006; Gέring et al., 2018). When used collaboratively, backcasting fosters shared exploration and learning, enabling participants to co-produce knowledge and design transition pathways (Carlsson-Kanyama et al., 2008).

Bibri and Krogstie (2020) emphasize the role of backcasting in guiding sustainability transitions, particularly within urban contexts. This approach connects long-term ambitions, such as decarbonization, equity, and resilience, with the decisions that need to be made in the short term. In doing so, it turns broad aspirations into a practical framework that supports innovation in public policy and planning.

When embedded in co-design workshops, backcasting becomes a powerful bridge between imagination and implementation. Citizens are not only invited to express their desires but also to map the concrete steps, stakeholders, and systems necessary to realize them. This process reinforces civic agency and repositions design as a tool for justice, strategy, and sustainability.

Building on these conceptual foundations, the following section presents two participatory design processes that sought to operationalize this intersection between co-design, futures thinking, and urban mobility. Conducted in Rio de Janeiro, these initiatives were structured around collaborative workshops and digital platforms and aimed to involve citizens not only in articulating their current mobility challenges but also in envisioning and shaping desirable mobility futures. The methodological approach combined speculative and strategic tools, including scenario-building and backcasting, to foster reflection, creativity, and agency among participants. By detailing how these processes were designed and implemented, the next section illustrates how citizen engagement can be mobilized as a critical resource for imagining and planning sustainable, inclusive, and context-sensitive urban mobility systems.

Methodological Approach: Designing with Citizens for Urban Mobility Futures

This section draws upon two empirical studies conducted in Rio de Janeiro that explored how citizens imagine and shape future urban mobility through participatory design. Both studies were based on design workshops and sought to activate citizen engagement not only through consultation, but through creative and situated participation. The approaches were conceived to explore two complementary dimensions of futures thinking: one oriented to free expression of desires and imaginaries, and the other to strategic construction of pathways to implementation. Both studies were submitted to the Ethics Committee of the author's home university, and all participants signed informed consent forms authorizing the use of their statements and co-created materials for academic research purposes.

Study 1: Envisioning Desirable Futures

To explore the needs, concerns, and aspirations of Rio de Janeiro's citizens regarding urban mobility, a series of co-design workshops were conducted between February and October 2021. Due to restrictions imposed by the COVID-19 pandemic, all sessions were held remotely using the Zoom video conferencing platform (Zoom.us). Participants were recruited through social media and direct emails and invited to take part in collaborative design sessions focused on short- and long-term futures. In total, six workshops were conducted, each lasting approximately two hours and bringing together 4–6 participants per session, adding up to 32 participants overall.

The group included 25 participants who identified as female and 7 as male, ranging in age from 18 to 69, and most between 18 and 39. Although participants lived and moved through different areas of the city, including the South, North, and West Zones, the majority were affiliated with a university and thus belonged to relatively privileged socioeconomic backgrounds. Still, as residents of a city marked by stark inequalities and shared public spaces, they were familiar with the dynamics and challenges of mobility in socially diverse environments. Participants' main transportation modes varied from car (10), subway (8), ride-hailing services (7), bus (4), and walking (3). Most were undergraduate and graduate students, alongside professionals from diverse fields.

After completing informed consent forms, participants received access to a shared Zoom meeting room and were introduced to the session's context through a short presentation of data, trends, and statistics on mobility in the city. Each participant was then guided to an individual board in the Miro platform (Miro.com), where they were invited to envision their ideal future for urban mobility in two timeframes: 2030 and 2050 (Figure 12.1). Participants were encouraged to express themselves using texts, drawings, or selections from a curated collection of pre-uploaded images on a collective board. To stimulate critical and imaginative reflection, each board included a set of guiding questions about their mobility expectations and values for the future (see Table 12.1.).

After an individual ideation period of about 20–30 minutes, participants shared their boards and reflections with the group. These presentations led to rich discussions about desired futures, personal experiences, and values associated with mobility in the city. This dialogical exchange allowed for both individual expression and collective sense-making.

The material generated during the workshops included the visual boards created by participants and the content of their verbal reflections, which were subsequently recorded and analyzed by the research team. An interpretive analysis was conducted using affinity diagramming (Marsh, 2018), initially organizing the wishes and concerns in a top-down structure based on a previously defined set of urban mobility criteria developed during an earlier phase of the research. These included traffic and congestion, punctuality and coverage of public transport, safety, fare integration, modal diversity, use of digital technologies, data sharing, costs and

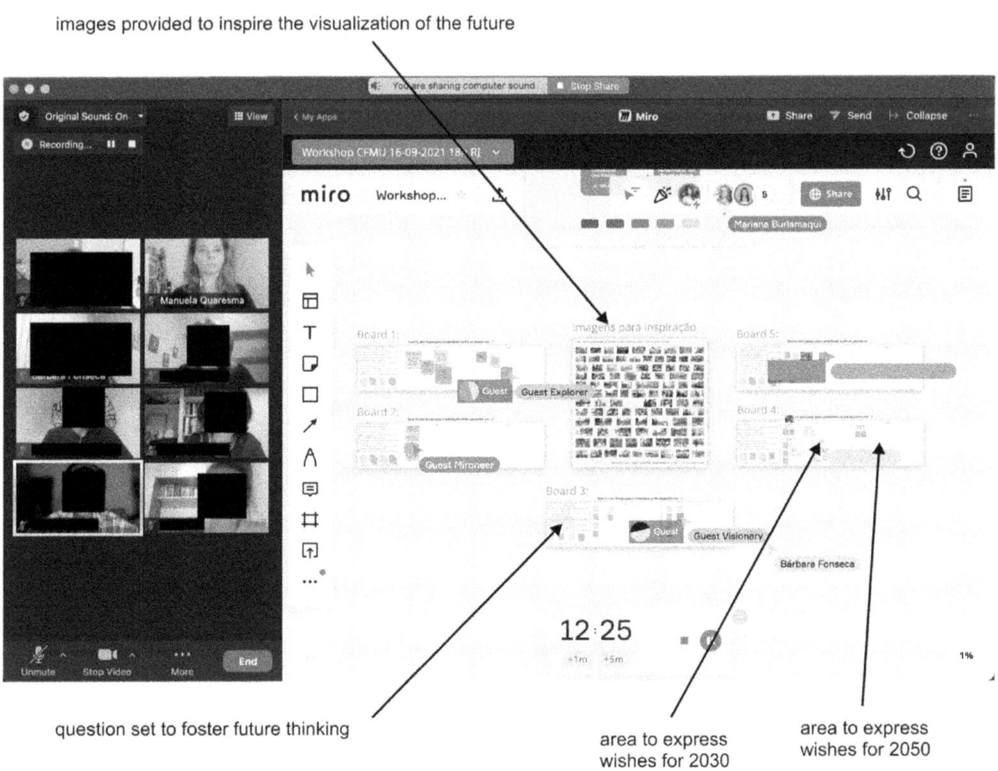

Figure 12.1 Workshop session setup in Zoom platform. Image by author.

Table 12.1 Question set used to foster future thinking during the workshop

Questions for future thinking on urban mobility
• *How would you envision mobility in your city given the population growth?* • *How do you envision yourself moving around the city in the best possible way? (consider what currently pleases you and what doesn't)* *What modes of transportation would you like to use in the future?* • *What do you think would be ideal for public transportation in your city? And for private transportation?* • *How much time do you think your commute should take in your city? (for work, shopping, etc.)* • *What technologies and services do you think would be ideal for your everyday life in your city in the future?* • *What new solutions could exist?* • *How do you think the ideal cost of transportation services should be?* • *How would you like a sustainable city to be?*

fares, environmental sustainability, travel time, accessibility, micromobility, shared mobility services, privacy, comfort, and vehicle ownership (Quaresma et al., 2022).

In a second layer of analysis, a bottom-up clustering was applied within each of these criteria to identify emerging themes across individual visions. This hybrid analytical process revealed recurrent desires and revealed which values were consistently prioritized across different groups, notably safety, sustainability, accessibility, autonomy, and affordability, while also shedding light on systemic gaps in the current mobility system.

Study 2: Exploring Pathways to Future Mobility through Backcasting

Building on the insights generated in the first study, the second phase of the research aimed to explore how the desirable futures envisioned by citizens could be transformed into actionable pathways. To do so, we adopted a backcasting approach structured around three macro-scenarios that had been previously presented in Quaresma et al. (2022). These macro-scenarios were the result of a literature review focused on identifying emerging trends and transformations in the field of urban mobility and were conceptualized as distinct yet complementary pathways for the evolution of mobility systems in the coming decades. The three macro-scenarios are:

1. ***Shared Mobility***, referring to services such as carpooling, vehicle sharing, ride-hailing services and collective transport systems;
2. ***Active Mobility***, emphasizing the use of human-powered transportation such as walking and cycling, and encouraging healthier, more sustainable urban practices;
3. ***High-Tech Mobility***, focusing on the implementation of advanced digital technologies, such as artificial intelligence, intelligent systems, and autonomous vehicles.

To deepen the understanding of how the desires expressed in Study 1 could be realized, eight mini-scenarios were selected for each macro-scenario, making a total of 24 mini-scenarios. These mini-scenarios were representative of the most recurrent and expressive citizen desires from the previous workshops. Each mini-scenario was described and illustrated with an AI-generated image to support participants' immersion and stimulate imaginative thinking.

Six new workshop sessions were conducted between April 2023 and March 2024, each focused on one of the three macro-scenarios, with two sessions per scenario. Each session involved four participants and was conducted online using the Zoom video conferencing platform. These participants were citizens from Rio de Janeiro but distinct from those who had taken part in the first study. The group included 15 participants who identified as female and 9 as male. Most were between 18 and 29 years old (18 participants), with a smaller number in the 30–39 (4 participants), 50–59 (1), and 60–69 (1) age brackets. Regarding transportation habits, their most frequently used modes included ride-hailing services (5), subway (4), car (4), bus (3), walking (2), bicycle (2), and BRT – Bus Rapid Transit (1). This group also reflected the urban diversity of Rio de Janeiro, with participants living in different regions of the city and exhibiting a range of commuting patterns and levels of access to transport infrastructure.

The workshops followed a collaborative backcasting method guided by the SQUID technique (Gray et al., 2011), designed to foster structured dialogue and speculative inquiry around possible futures. At the beginning of each session, participants were welcomed with a warm-up exercise and introduced to the selected macro-scenario and its related trends. The SQUID technique was explained through a practical example before engaging with the actual mini-scenarios. In the main dynamic, participants interacted with the Miro platform (Figure 12.2), where each mini-scenario was presented.

The eight mini-scenarios were explored in two sequential stages during the workshop. In the first stage, participants interacted with four mini-scenarios, applying the SQUID technique (Gray et al., 2011) through two structured cycles. In each cycle, a participant would first formulate concerns by writing down questions about the feasibility or challenges of achieving a specific mini-scenario. In the next round, another participant—who had not previously formulated the concerns—would read those questions and provide answers, understood as possible solutions or interventions to address the issues raised. This procedure was repeated across the four initial mini-scenarios.

In the second stage, the same process was applied to the remaining four mini-scenarios, again involving two full cycles of concern-generation and solution-response. This structure ensured that all participants alternated between identifying barriers and imagining actionable strategies,

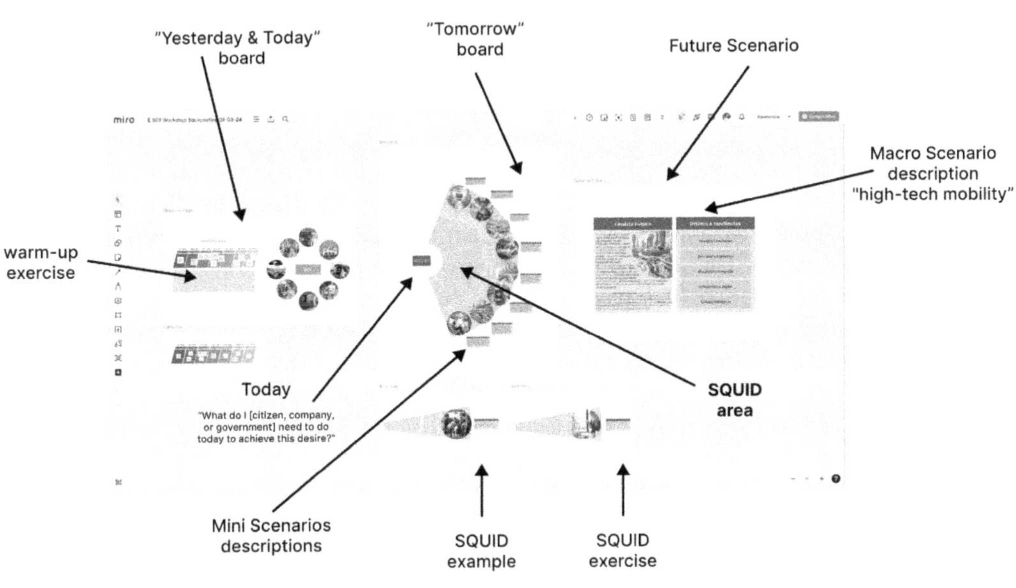

Figure 12.2 Miro board used to conduct the backcasting method.

fostering both critical reflection and collective creativity. The result was a rich set of citizen-generated insights about the challenges and opportunities involved in transitioning toward the desirable futures outlined in each scenario.

To analyze the resulting data, researchers extracted all questions and responses generated during the sessions. The material was then subjected to an affinity diagramming analysis (Hanington & Martin, 2019), grouping ideas into thematic clusters. These clusters were organized into four major categories:

1. *Public Policies*—including laws, regulations, incentives, and institutional changes proposed to enable the mini-scenarios;
2. *Products*—representing new or improved physical artifacts (e.g., vehicles, infrastructure) aligned with the visions of the future;
3. *Services*—such as travel planning apps, shared systems, or new models of urban service provision;
4. *Product-Service Systems (PSS)*—integrated ecosystems combining physical and digital components to deliver comprehensive mobility solutions.

The structured use of the SQUID methodology within a backcasting framework allowed for a nuanced understanding of both aspirational futures and the pathways necessary to realize them. It also ensured that the proposed transformations were not solely abstract or utopian, but anchored in concerns, constraints, and actionable ideas articulated by the citizens themselves.

Citizen Engagement and Participation in Practice

The two studies presented in this chapter were designed to explore how citizens in Rio de Janeiro envision desirable futures for urban mobility and identify potential pathways to achieve them. Although they followed different methods, both prioritized direct engagement with citizens through participatory workshops conducted remotely, in response to the constraints imposed by the COVID-19 pandemic. While in-person workshops often allow for greater interpersonal interaction, the use of online platforms with interactive tools proved to be highly engaging. Participants were motivated by the opportunity to contribute to a topic that directly affects their daily lives, and the structure of the activities helped sustain focus and participation throughout the sessions.

The first study focused on capturing participants' desires and aspirations for mobility in the years 2030 and 2050. Through visual and reflective exercises facilitated by digital platforms, participants were invited to imagine ideal futures, expressing their values and concerns in creative and personal ways. The methodology encouraged subjective and symbolic expression, allowing for the construction of future narratives rich in meaning. The digital format also enabled the inclusion of participants from different regions of the city, broadening the diversity of perspectives. Notably, participants demonstrated high levels of engagement and willingness to share experiences, which were often marked by the complexities of navigating the city. Their imaginative contributions revealed both frustrations with the present and ambitions for more inclusive and sustainable mobility futures.

The second study built upon the mapped desires from the first, aiming to investigate more structured possibilities for achieving those visions. The sessions applied a backcasting method supported by the SQUID technique, guiding participants through more specific scenario explorations and encouraging critical reflection on challenges and actionable solutions. This technique was selected after a series of preliminary tests with different activity models, conducted to ensure that the remote format would foster meaningful dialogue and avoid repetitive expressions of desire. The SQUID technique proved particularly effective in prompting participants to shift from aspiration to strategy, offering concrete proposals despite the acknowledged political, economic, and technological constraints faced by citizens in Rio de Janeiro. Unlike the first study, which was more open-ended and exploratory, this second approach required participants to adopt a more analytical and solution-oriented stance. Nonetheless, the emphasis remained on citizen engagement, with a focus on guiding them toward identifying feasible strategies for change.

It is important to note that the participants in the two studies were not the same. The first group contributed imaginative visions and personal mobility experiences, while the second group was specifically mobilized to explore pathways toward the previously identified desirable futures. Both studies were conducted online using tools like Zoom and Miro, ensuring continuity in participatory engagement and enabling a rich, collaborative visual experience.

The comparison between the two workshop formats reveals complementary ways of engaging citizens in the construction of future mobility scenarios. The first created a space for listening and collective imagination, which was essential for capturing legitimate and diverse societal desires. The second focused on developing realistic strategies through a more systematic process of reflection, supported by the SQUID technique. The originality of this work lies in combining these two approaches, which are often applied in isolation, into a coherent and sequential process where the second phase directly builds on the outcomes of the first. This design not only preserved the open and creative nature of co-design but also introduced a targeted mechanism to guide participants toward structured and goal-oriented reasoning. By fostering a more strategic mindset and linking aspirational visions to actionable pathways, the methodology expanded the potential of participatory workshops to deliver concrete, context-specific outcomes while sustaining meaningful citizen engagement.

In this sense, participants were not only contributors of experiences and needs but were engaged as **strategic co-thinkers**, actively reflecting on how current challenges could be connected to future aspirations. This positioning shifted the role of the workshops from a conventional solution-oriented exercise to a space for strategic anticipation, where citizens could imagine desirable futures and outline the conditions for making them possible.

Citizen-Generated Visions and Strategic Proposals

Key Aspirations for Urban Mobility

The first round of workshops revealed a rich and multifaceted set of aspirations among Rio de Janeiro's citizens concerning the future of urban mobility. These aspirations were organized according to two future timeframes (2030 and 2050) and clustered through an analytical process that combined top-down classification (based on pre-existing urban mobility criteria) with bottom-up identification of emergent themes.

For the 2030 horizon, participants envisioned a city that was more accessible, equitable, and efficient in its transportation systems. Central to their aspirations were improvements in the availability, integration, and punctuality of public transport. Many citizens expressed a desire for decentralized mobility options that would allow people to move more freely without relying on car-centric infrastructures. Concepts such as "proximity-based cities," neighborhood mobility hubs, and integrated modal systems frequently appeared across different participant boards. Sustainability, affordability, and safety, especially for women, children, and people with disabilities, were also prominent themes for the near future.

Looking further ahead to 2050, participants projected more radical and transformative scenarios. Many imagined a city where cutting-edge technologies, such as autonomous vehicles, aerial transportation, and advanced navigation systems, would be embedded into daily life. However, these technological aspirations were rarely presented in isolation; rather, they were linked to broader ideas of collective well-being, community cohesion, and environmental responsibility. Several participants envisioned micro-communities with strong local ties and low-impact transportation solutions, suggesting a desire for innovation grounded in social and ecological values.

Across both timeframes, participants placed strong emphasis on reducing inequalities in mobility access and ensuring that future systems are inclusive and participatory by design. There was a clear preference for public over private solutions, with many boards advocating for secure shared mobility services, expanded bike and pedestrian infrastructure, and seamless digital integration across transportation modes. While 2030 visions tended to prioritize urgent and pragmatic changes, the 2050 scenarios invited more speculative, imaginative responses, yet both reflected a consistent call for a mobility system that is humane, intelligent, and fair.

Solutions Identified through Backcasting

The second series of workshops, based on a backcasting method, aimed to transform the aspirations identified in the first study into actionable strategies. Through structured sessions applying the SQUID technique, citizens worked collaboratively to explore challenges and propose concrete solutions for achieving the desirable mobility futures previously envisioned. Each session focused on one of three macro-scenarios, composed of eight mini-scenarios derived from the most recurrent and meaningful citizen desires identified in the first study.

Over the course of the six workshops, participants collectively generated a total of 397 distinct solution proposals. These contributions were the result of carefully facilitated activities that encouraged critical reflection and creativity while drawing on participants' daily experiences with the city's transportation system. The high volume of contributions, considering the limited number of participants, demonstrates the effectiveness of well-designed participatory methods in eliciting motivated and meaningful engagement from citizens.

Participants' proposals were categorized into four types: public policies, services, products, and product-service systems. Notably, public policies and service-based solutions dominated the responses, indicating a strong civic orientation and a belief in the role of collective infrastructure and governance in shaping future mobility systems.

Among the policy proposals were ideas such as tax incentives for sustainable mobility options, subsidies for low-income users, and new regulations to integrate multiple transport systems into a unified platform. Participants also suggested the creation of participatory councils to ensure user feedback is continuously incorporated into urban planning processes. In the service domain, solutions included app-based systems that combine real-time data from multiple modes of transport, community-run bike repair stations, and inclusive ride-sharing networks designed specifically for underserved neighborhoods.

Several participants envisioned hybrid solutions that blended services and products, such as a mobility subscription model that combines access to metros, buses, bike-sharing, and micromobility devices within a single, digitally integrated payment and tracking system. Other solutions emphasized the importance of maintenance and education, for example, programs to teach cycling skills in schools or community workshops on urban mobility rights and digital platforms.

While the participants were free to explore a wide range of ideas, it is important to acknowledge a limitation of the process: the workshops were not attended by public officials or representatives from government agencies. This absence may have encouraged more imaginative and unrestricted thinking, but it also meant that proposed solutions were not immediately negotiated against institutional or bureaucratic constraints. This points to an opportunity and a challenge for future engagements that aim to bridge citizen input and public policy development.

By alternating between identifying barriers and proposing responses, the backcasting process enabled participants to think critically and constructively about implementation pathways. The insights generated in these sessions illustrate how citizen engagement, when properly scaffolded, can produce both visionary and grounded ideas for systemic change. Rather than remaining at the level of aspiration, the second study offered a collective mapping of what needs to be in place, politically, economically, and socially, to realize the desirable futures articulated in the first phase.

Concluding Remarks and Future Directions

The dual-stage methodology presented in this chapter demonstrates how the integration of participatory foresight approaches (speculative future visioning and collaborative backcasting) can support citizen-centered innovation in urban mobility. The first stage captured participants' aspirations and everyday frustrations, illuminating consistent values such as safety, integration, accessibility, and sustainability. The second stage translated these aspirations into actionable strategies, generating 397 concrete proposals across six workshops. This outcome illustrates the potential of structured, future-oriented engagement to activate creative, yet pragmatic, citizen participation in the design of public services. The novelty of this approach lies in reframing the role of design workshops: rather than being limited to solution-generation, they were structured to engage citizens as strategic co-thinkers, capable of linking present challenges with long-term visions and of proposing concrete pathways toward desirable futures.

From a methodological perspective, the use of design tools such as visual platforms (Miro) and the SQUID technique enabled high levels of engagement, even in a remote format. Despite concerns that digital environments might limit interaction, the workshops facilitated motivation and active participation, likely due to the personal relevance of urban mobility in participants' daily lives. However, one limitation of the process was the absence of institutional actors. While this enabled more free-flowing citizen proposals, it also distanced the sessions from real-world governance constraints. Addressing this gap in future iterations by including public managers, funders, and technology providers may help assess the feasibility of the proposed ideas and foster co-responsibility for implementation.

The results of this research offer strategic contributions for public innovation, but practical implementation is not without challenges. Political resistance, funding limitations, and infrastructural barriers can all inhibit the realization of citizen-proposed solutions. In this sense, future research could focus on prototyping selected services and testing them in collaboration with stakeholders to understand not only their desirability but also their viability and feasibility. Moreover, although the methodology proved effective in Rio de Janeiro, its broader applicability remains to be tested. The framework could be replicated in other urban contexts or in other areas of public services beyond mobility. Comparative studies across cities would help validate and refine the approach, offering insight into its scalability and adaptability.

In summary, this chapter shows how engaging citizens in the design of future mobility systems, through a structured, strategic, and imaginative process, can generate rich and diverse insights. It advances a replicable model for participatory engagement that bridges emotional and visionary thinking with actionable, system-oriented proposals. As a contribution to citizen-centered design in the Global South, it highlights the importance of grounding innovation in people's lived experiences while pointing toward future pathways of prototyping, scaling, and institutionalizing solutions co-developed with the public.

Acknowledgments

This study was funded in part by the Conselho Nacional de Desenvolvimento Científico e Tecnológico (CNPq) and by the Fundação Carlos Chagas Filho de Amparo à Pesquisa do Estado do Rio de Janeiro (FAPERJ). I would also like to express my sincere gratitude to the undergraduate students and research scholarship holders who actively contributed to the development of this project. This chapter was revised with the support of artificial intelligence tools, while its conceptual development and authorship remain entirely human.

References

Bibri, S. E., & Krogstie, J. (2020). Data-Driven Smart Sustainable Cities of the Future: A Novel Model of Urbanism and Its Core Dimensions, Strategies, and Solutions. *Journal of Futures Studies*, *25*(2), 77–94. https://doi.org/10.6531/JFS.202012_25(2).0009

Borba, F., Dantas, A. C., & Dutt-Ross, S. (2022). *Mobilidade Urbana no Rio de Janeiro*. Editora CRV. www.mobilize.org.br/estudos/486/mobilidade-urbana-no-rio-de-janeiro--diagnostico-a-partir-da-percepcao-do-cidadao.html

Carlsson-Kanyama, A., Dreborg, K. H., Moll, H. C., & Padovan, D. (2008). Participative Backcasting: A Tool for Involving Stakeholders in Local Sustainability Planning. *Futures*, *40*(1), 34–46. https://doi.org/10.1016/j.futures.2007.06.001

European Commission. (2020). Smart Cities. https://ec.europa.eu/info/eu-regional-and-urban-development/topics/cities-and-urban-development/city-initiatives/smart-cities_en

Gebauer, H., Johnson, M., & Enquist, B. (2010). Value Co-Creation as a Determinant of Success in Public Transport Services. *Managing Service Quality: An International Journal*, *20*(6), 511–530. https://doi.org/10.1108/09604521011092866

Géring, Z., Király, G., Csillag, S., Kováts, G., Köves, A., & Gáspár, T. (2018). Vision(s) of the University. Applying Participatory Backcasting to Study the Future of Higher Education. *Journal of Futures Studies*, *22*(4), 61–82. https://doi.org/10.6531/JFS.201806.22(4).0005

Gidley, J. M. (2017). *The Future: A Very Short Introduction*. Oxford University Press.

Gray, D., Brown, S., & Macanufo, J. (2011). *Gamestorming: A Playbook for Innovators, Rulebreakers, and Changemakers*. O'Reilly Media.

Hanington, B., & Martin, B. (2019). *Universal Methods of Design Expanded and Revised: 125 Ways to Research Complex Problems, Develop Innovative Ideas, and Design Effective Solutions (Revised)*. Rockport Publishers. www.amazon.com.br/gp/product/1631597485?ref=em_1p_1_ti&ref_=pe_1249810_506473820

Khan, Z., Dambruch, J., Peters-Anders, J., Sackl, A., Strasser, A., Fröhlich, P., Templer, S., & Soomro, K. (2017). Developing Knowledge-Based Citizen Participation Platform to Support Smart City Decision Making: The Smarticipate Case Study. *Information*, *8*(2), 47. https://doi.org/10.3390/info8020047

Kirimtat, A., Krejcar, O., Kertesz, A., & Tasgetiren, M. F. (2020). Future Trends and Current State of Smart City Concepts: A Survey. *IEEE Access*, *8*, 86448–86467. https://doi.org/10.1109/ACCESS.2020.2992441

Kudo, H. (2016). Co-design, Co-creation, and Co-production of Smart Mobility System. In P. Rau (Ed.), *Cross-Cultural Design. CCD 2016. Lecture Notes in Computer Science* (pp. 551–562). Springer. https://doi.org/10.1007/978-3-319-40093-8_54

Marsh, S. (2018). *User Research: A Practical Guide to Designing Better Products and Services*. Kogan Page.

Mattelmäki, T., Brandt, E., & Vaajakallio, K. (2011). On Designing Open-Ended Interpretations for Collaborative Design Exploration. *CoDesign*, *7*(2), 79–93. https://doi.org/10.1080/15710882.2011.609891

Moreno, C. (2020). *Vie urbaine et proximité à l'heure du Covid-19*. Humensis. https://ideas.repec.org/p/hal/journl/hal-03259768.html

Morozov, E., & Bria, F. (2019). *A cidade inteligente – Tecnologias urbanas e democracia*. Ubu Editora.

National Association of City Transportation Officials. (2018). Global Street Design Guide. https://nacto.org/publication/global-street-design-guide/

Noring, L., Fróes, I. C. G., & Tellgren, D. (2018). Contextualising Mobility Variables. *WIT Transactions on Ecology and the Environment*, *217*, 165–173. https://doi.org/10.2495/SDP180161

Organization for Economic Co-operation and Development. (2020). Smart Cities and Inclusive Growth. www.oecd.org/cfe/cities/OECD_Policy_Paper_Smart_Cities_and_Inclusive_Growth.pdf

Quaresma, M., Fonseca, B. N. da, & Burlamaqui, M. (2022). Jornadas Futuras Para a Mobilidade Urbana em Cidades Inteligentes. *Blucher Design Proceedings*, 952–965. https://doi.org/10.5151/18ergodesignusihc2022-61

Quist, J., & Vergragt, P. (2006). Past and Future of Backcasting: The Shift to Stakeholder Participation and a Proposal for a Methodological Framework. *Futures*, *38*(9), 1027–1045. https://doi.org/10.1016/j.futures.2006.02.010

Sanders, E. B.-N., & Stappers, P. J. (2008). Co-Creation and the New Landscapes of Design. *CoDesign*, *4*(1), 5–18. https://doi.org/10.1080/15710880701875068

Shaheen, S., Cohen, A., Zohdy, I., & Kock, B. (2016). Shared Mobility: Current Practices and Guiding Principles Brief. https://ops.fhwa.dot.gov/publications/fhwahop16022/fhwahop16022.pdf

United Nations Department of Economic and Social Affairs (UNDESA). (2018). *World Urbanization Prospects Revision: The 2018 (ST/ESA/SER.A/420)*. https://population.un.org/wup/Publications/Files/WUP2018-Report.pdf

Vallet, F., Puchinger, J., Millonig, A., Lamé, G., & Nicolaï, I. (2020). Tangible Futures: Combining Scenario Thinking and Personas – A Pilot Study on Urban Mobility. *Futures*, *117*, 102513. https://doi.org/10.1016/J.FUTURES.2020.102513

13 Engaging Citizens in Robot Dog Interaction in Smart Cities

Perceptions in Public Spaces

Emma Kirjavainen, Matilda Kalving, Siiri Paananen, Ashley Colley, and Jonna Häkkilä

Introduction

Rapid technological development and digitalization have reshaped society, embedding mobile and ubiquitous technologies into everyday urban life. Autonomous cars and smart buildings illustrate how cities are integrating technology into public infrastructure. This trend is expected to intensify as artificial intelligence (AI) and autonomous systems become more deeply embedded in urban environments. In this paper, we use the term "smart cities" to refer to urban contexts where AI-enabled technologies support context-aware services and interactions with citizens. Social robots are among the emerging technologies expected to play a role in this transformation (Henschel et al., 2021).

The role of robots in society is shifting. Once confined to factory production lines, robots are now present in public and social domains – delivering food (Martinez et al., 2023; Väätäjä et al., 2023), enabling telepresence at events (Neustaedter et al., 2016), supporting healthcare (McNelis et al., 2012), contributing to education (Yadollahi et al., 2022), and assisting in customer service (Xiao & Kumar, 2021). As robots move into everyday settings, their acceptance and social integration increasingly depend on cultural norms and expectations. These robots are expected to not only perform practical tasks but also to align with everyday practices and elicit social and emotional responses (Breazeal, 2004a; Darling, 2016).

This study focuses on public perceptions of quadrupedal robot dogs in everyday urban settings. We investigate how people respond to such robots in public spaces and what concerns or expectations arise as these machines become more common in civic life. The robot used in this study is the widely available Unitree Go1 (Unitree, n.d.) (Figure 13.1), a low-cost quadruped capable of outdoor mobility and carrying small payloads. It measures 645 × 280 × 400 mm, weighs 12 kg, and can move at speeds of up to 17 km/h. The robot realistically mimics the size and agility of a medium-sized dog. Unlike real dogs, already familiar in public roles such as guiding or security, robot dogs raise distinct questions about autonomy, intent, and technological presence. Our interest lies not in animal behavior but in how the zoomorphic form shapes public perceptions of an autonomous machine operating in shared space.

To explore this, we outline a framework for understanding how robot dogs are perceived in public and social environments. Such understanding can support the design of future robotic systems that align with societal values and expectations. The study offers insights for designers and engineers into how form, context, and behavior influence acceptance and imagined roles.

We used a mixed-methods approach combining an online survey and in-situ field interviews to investigate perceptions of robot dogs in real-world contexts. The survey yielded responses from a broad sample, while interviews conducted in public locations provided in-depth insights into individual reactions and interpretations (Millar & Tracey, 2009). Together, these methods

DOI: 10.4324/9781003667872-17

Figure 13.1 Examples of the contexts where the acceptance of robot dogs was explored in the field study.

offered a more complete understanding of public interaction with robot dogs than either method could alone.

Related Work

Interacting with Robots in Social Spaces

Breazeal (2004b) provides a foundational perspective on social interactions in HRI, emphasizing the robot's role in facilitating engagement. Social robots are often de- signed to communicate through social cues and exhibit adaptable learning behavior (Darling, 2016). They may also use AI and machine learning to recognize and respond to human actions (Breazeal, 2004a).

Leite et al. (2013) reviewed long-term studies involving robots in domains such as healthcare, education, homes, workspaces, and public areas. They emphasized the specific challenges of deploying robots in public spaces, particu- larly regarding adaptability to unpredictable environments. Abrams et al. (2021) introduced the concept of "Existence Acceptance" to capture how people respond to delivery robots in incidental encounters, highlighting differences from traditional technology acceptance models.

Long-term studies in public spaces further explore how robots are perceived over time. Niemelä et al. (2017) studied a service robot in a shopping mall, while Sung et al. (2009) and Moshkina et al. (2014) examined sustained engagement in other public settings. Research has also extended to interactions between robots and animals: Morovitz et al. (2017) found that dogs preferred robots without human-like movements or sounds, suggesting that animal-friendly robot design may differ from human-oriented design. Building on this, Väätäjä et al. (2023) investigated how dogs respond to delivery robots in urban settings. While most dogs reacted neutrally, a minority showed avoidant (8%) or hostile (9%) responses, revealing important edge cases for public deployment in pet-populated areas.

Zoomorphic Robots

In HRI, zoomorphism refers to the design of robots with animal-like characteristics. Löffler et al. (2020) studied the uncanny valley in zoomorphic robots, showing that likability follows a U-shaped curve: robots are less likable when they appear ambiguously between machine and animal, but more likable when either clearly robotic or convincingly animal-like. In line with this, Sandry (2015) argue that effective interaction does not require animal features, as machine-like robots can also perform social functions. Motion cues are another benefit of zoomorphism: Walker (2018) showed that animal-like movement can make a robot's intent more easily understandable to people, though they also explored ways to convey motion intent in non-zoomorphic robots. Hegel et al. (2009) linked the robot appearance to functional expectations. Human-like robots were associated with roles in security, healthcare, and teaching, while animal-like robots were expected to serve as companions, toys, or pets. A cross-cultural study by Li et al. (2010) supported this, noting zoomorphic robots are particularly suitable for entertainment purposes.

Robot dogs represent a specific zoomorphic form factor that has been explored in various contexts. Studies using Sony's AIBO include Kahn Jr et al. (2004), who examined preschoolers' interactions with the robot dog, finding parallels with real dog behavior. Other research has compared AIBO to real dogs (Melson et al., 2005), investigated autonomy vs. teleoperation (Chernyak & Gary, 2019), and examined emotional attachment (Weiss et al., 2009). Studies conducted outdoors with robot dogs have investigated people's interaction and perception with them. Wu et al. (2023) explored how a robot dog could support elderly people walking outdoors, reporting positive reactions and natural interactions. Paananen et al. (2023) recorded and studied 192 interactions with a robot dog at a public event, finding that 17% involved gestures typically used when greeting or petting a real dog.

Together, these studies demonstrate that robot dogs elicit social responses based on their form, but they also reveal that most prior work has occurred in controlled environments or with toy-like robots.

Our Contribution

While previous research has explored human interaction with robot dogs – often in private or experimental settings – our study makes two key contributions:

- **Public contexts:** Most studies on robot dogs and HRI have been conducted indoors or in lab environments. Our study extends this by investigating how people respond to robot dogs in real public contexts, offering a more realistic perspective on social interaction and acceptance.
- **Biomimetic robot dog:** Much prior work used toy-like robots, such as Sony's AIBO. We focus instead on a robot dog that mimics the size and movement of a real dog, enabling new kinds of physical interaction and use cases in public spaces. While AIBO emphasizes emotional play (e.g., tail wagging), our study explores practical and social expectations of robot dogs performing tasks in shared civic spaces.

This work advances the understanding of zoomorphic public robots by examining how physical design, context, and cultural expectations intersect in everyday interactions.

Online Survey

Through an online survey, we aimed to contribute large-sample data to our overall dataset. As with the other study modes, the main goal of the online survey was to understand public perception regarding the presence and activities of robot dogs in various public settings.

Study Set-up

The survey collected participants' insights on five scenarios: a shopping mall, a museum, a countryside path, a train station, and a beach. The contexts for the scenarios were selected based on Koelle et al. (2020), which presents the common study locations used in studies on social acceptability for HCI systems. These locations also reflect typical public environments where smart city technologies are deployed or envisioned, such as transport hubs and managed urban spaces. Each scenario was presented with a probe image of a robot dog in the scenario (Figure 13.2). For each scenario, participants rated the acceptability of a robot dog performing work tasks at the location on a 7-point rating scale, ranging from 1 (not at all acceptable) to 7 (very acceptable). Participants then provided their ideas on the possible duties of the robot dog at the location with a free-text response. After reviewing all the scenarios, participants gave feedback on their perception of the general problems they perceived with robot dogs working in public places.

The survey was distributed through Prolific (www.prolific.co/). Participants were selected as native English speakers located in the UK who had a task approval rate of above 95%. The survey took a median time of 4 m 5 s to complete, for which participants were compensated €0.70. The compensation level was suggested by the recruitment platform and corresponded to an hourly rate of approximately €10. A total of 100 participants completed the survey, 59 identifying as women and 41 as men. Participants ranged from 20 to 79 years of age, with a mean age of 40.2 years (SD = 13.7).

Results

Rating Scale Responses

Participants' responses to the rating scale questions are shown in Figure 13.3. The results reveal a broad spread of opinions across all of the usage scenarios. Overall, 50% of the responses were on the negative side, with 21% of respondents reporting the strongest negative rating, '1 not at all'. Altogether 9% of respondents were strongly positive (rating '7 very much') about the use of robot dogs. Interestingly, participants did not appear to have a general liking or dislike of robot dogs, with only 17% giving consistently negative ratings, below the scale midpoint, and 6% giving consistently positive ratings. This suggests that, on an individual level, the usage scenario is an important factor in the acceptance of robot dogs.

Due to violations of the normality assumption, non-parametric tests were employed for the analysis. A Kruskal-Wallis test revealed a statistically significant difference in ratings across the five scenarios, $H(4) = 25.27$, $p < 0.001$. Post-hoc Mann-Whitney pairwise comparisons between the scenarios were conducted. A Bonferroni-corrected alpha level of $\alpha = 0.005$ was applied to control for Type I errors due to multiple comparisons. Significant differences were found between the following pairs of scenarios:

- Shopping Mall vs. Museum: $U = 6172.5$, $p = 0.0038$
- Shopping Mall vs. Beach: $U = 6457.0$, $p = 0.0003$
- Countryside Path vs. Museum: $U = 3622.0$, $p = 0.0007$
- Countryside Path vs. Beach: $U = 6581.0$, $p = 0.00009$

No significant differences were observed among the remaining pairwise comparisons. The results indicate that participants' acceptance of robot dogs working in the shopping mall context

Figure 13.2 Probe images used in the online survey. From top left: Shopping mall, museum, countryside path, train station, beach.

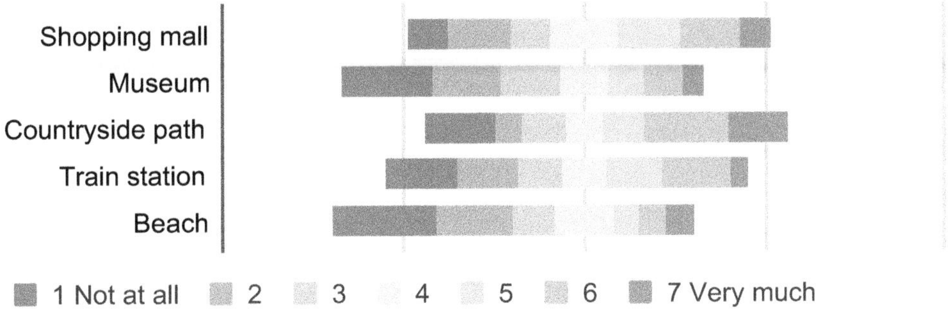

Figure 13.3 Results from the online survey rating scale question on acceptance of a robot dog working in each context ($n = 100$).

was significantly higher than in the Museum and Beach scenarios. Similarly, acceptance in the Countryside Path scenario was significantly higher than in the Museum and Beach scenarios.

Qualitative Analysis

Participants' free-text responses to the open questions were analyzed using an open coding approach (Strauss, 1987): One researcher defined the codebook and made an initial coding for the dataset. Using the codebook, a second researcher then independently coded the dataset. The researchers then discussed the differences in their coding and produced a final coded dataset. In the discussion phase, several of the codes were merged. The top five codes identified in each scenario are shown in Table 13.1. In the following, we discuss the most prominent themes in more detail.

Surveillance and Security. The Surveillance and Security theme refers to participants envisioning robot dogs monitoring or ensuring safety in public spaces. Overall, the task of Surveillance and Security was most frequently cited, with 84 mentions, and was the highest-mentioned code in the museum (28 mentions) and beach (19 mentions) scenarios. One participant noted the robot dog's utility at the beach, stating, "Alert when people may need aid/ assistance, or alternatively when inappropriate behaviour is occurring" (Participant P18). This reflects engagement through persuasion – a passive but important role in public trust.

Localization and Guidance. Participants frequently imagined the robot dog helping locate areas, people, or items — particularly in the shopping mall and countryside scenarios, e.g., "It could help me find the food court quickly" (P12), and "Guide people on walking trails and give directions" (P46). Here, engagement is assistive and functional, helping users navigate unfamiliar or complex environments. In the shopping mall, the task of Localization – referring to locating places and products – gained the most mentions (25 mentions) and was the second-highest-mentioned task overall (61 mentions).

Cleaning and Maintenance. The third most commonly cited task was Cleaning (56 mentions), which, although not ranked in the top three tasks in any specific location, consistently appeared among the highest-priority tasks for a robot dog across all lo- cations. Comments included, e.g.,

Table 13.1 The top five mentioned task codes in each scenario

Shopping mall	*Museum*	*Countryside path*	*Train station*	*Beach*
Localization (25)	Surveillance and security (28)	Tour guide (21)	Alerts (11)	Surveillance and security (19)
Goods Transportation and Delivery (25)	Tour guide (18)	Localization (18)	Information (10)	Cleaning Services (16)
Surveillance and Security (22)	Information (11)	Surveillance and security (13)	Localization (9)	Goods Transportation and Delivery (15)
Cleaning Services (19)	Cleaning Services (9)	Communication to a third party (11)	Crowd Management (5)	Alerts (10)
Customer Engagement (12)	Localization (6)	Cleaning Services (10)	Space Limitations (3)	Vendor functions (10)

Note: The bracketed number is the number of mentions of the code.

"It could pick up small trash or clean high-traffic areas" (P21) and "Would be useful in the mall to alert staff or clean messes" (P49). This reflects indirect engagement contributing to a shared public good.

Emotional Support and Accessibility. Participants also proposed that the robot dog could fulfill roles related to accessibility, specifically in the areas of Mobility and Visual Support (16 mentions) and Emotional Support and Companionship (16 mentions). As suggested by one participant, "Help guide visually impaired people so they can shop on their own" (P4). This emphasizes emotional and inclusive engagement, expanding the role of robots beyond utility into empathetic presence.

Negative Themes. Concerns and Limitations Participants also raised concerns, which reflect barriers to acceptance and engagement. Negative feedback on the potential of robot dogs was categorized under several themes. Denial (62 mentions), which signifies skepticism about the robot dog's utility or appropriateness for certain locations, was most frequently mentioned in the general questions (15 mentions), at the beach (14 mentions), and at the museum (12 mentions). Practicality was a common reason for this skepticism. For example, one participant (P19) remarked about train stations, "None – tube stations are too crowded as it is", while another questioned the robot dog's utility at the beach, stating, "The sand would make movement difficult. I can't see how it would be helpful" (P74). Uncertainty or Unawareness (74 mentions) about the robot dog's possible roles was another prevalent theme, particularly at museums (18 mentions), beaches (12 mentions), and shopping malls (15 mentions). Other concerns included Vandalism (19 mentions) and Environmental limitations (22 mentions), such as Weather conditions (16 mentions) and spatial constraints (6 mentions). Participants also expressed worries about the robot dog causing Space limitations (40 mentions) for observers, posing trip hazards, and being Annoying (28 mentions). One participant summarized multiple concerns about the beach scenario, stating, "[It would] struggle to move on the sand, get wet, get stolen, and be played with by kids. I would feel spied on. It could potentially handle security, but a dog-shaped robot is just a robot" (P52).

These qualitative findings show that people evaluate robots not only based on technical prowess but also on perceived roles, emotional presence, and social context. Participants imagined robot dogs not only as tools but also as part of public life – as guides, helpers, companions, and sometimes as intruders. This challenges traditional notions of interaction as mere interaction and moves toward a more holistic understanding that includes emotional resonance, situational appropriateness, and cultural expectations. These findings offer a distinct contrast to typical lab-based HRI studies, where participants are often primed to consider robot functionality in abstract or constrained terms. Here, participants responded based on real-world situational relevance, leading to emergent concerns such as physical space limitations, bystander reactions, and environmental obstacles — aspects less commonly surfaced in controlled environments.

Field Interviews

Through field interviews, we aimed to gain an in-depth understanding of individual viewpoints on the perception of robot dogs. By conducting the interviews in the actual usage contexts, participants' perceptions are based on the actual situation rather than an imagined case, and critical issues that are not apparent in lab-based testing may be identified.

Study Set-up

In the field study, the robot dog was taken to six different locations: a museum, a shopping mall, the town square, a country hiking path, a beach, and a university campus area (Figure 13.4). One researcher controlled the robot dog and acted as an observer from a short distance away from the dog. A second researcher walked alongside the dog and engaged with passers-by. Those passers-by who showed interest in the robot dog were invited to take part in an interview. Each interview lasted 10–15 minutes and was audio-recorded for later analysis. Altogether, 21 participants (12 identifying as women and nine as men) participated in the study. Participants' ages ranged between 20 and 70 years, with a mean age of 44.

After completing a consent form, the interview included the introduction of the robot dog and a background questionnaire, which included previous experience with various robots, e.g., robot vacuum cleaners. Participants were then questioned on their acceptance of a robot dog in the current location and the perceived challenges. Participants were then asked to suggest other contexts where a robot dog would be useful or unwelcome. As a final task, participants completed the Robotic Social Attribute Scale (RoSAS) (Carpinella et al., 2017) based on the robot dog in the current location. This scale contains 18 attributes, with which the participants rated the robot dog on a scale of 1 to 5.

Results

Unlike lab studies, the public settings allowed spontaneous interactions with a diverse range of people, including those who would not normally opt into robot research, offering insights into unanticipated responses and practical constraints. In each interview location, the robot dog successfully captured the attention of numerous passers-by, although converting that interest into active participation in the interviews was challenging. Overall, the robot dog was met with a largely positive reception in all the locations. Many people approached the researchers to inquire about the robot and often expressed that they found it endearing. However, there were some exceptions, such as a child who initially reacted to the robot with fear. With the encouragement

Figure 13.4 The six interview locations. From top left: museum, shopping mall, town square, forest path, beach, and university campus.

of her parents and our gentle introduction of the robot dog, she was eventually able to overcome her apprehension.

The majority of interviewees (20/21 participants) felt that the robot dog was well-suited for the location where the interview took place. However, participants considered that the robot dog's specific function was a critical factor in its appropriateness for any setting. When discussing potential roles for the robot dog, transporting items was the most commonly suggested task (9/21). Other tasks, such as navigation guidance (8/21), disseminating information (2/21), and cleaning tasks (2/21) were also mentioned. Specific assistance tasks, such as opening doors, were highlighted in interviews conducted in the indoor locations. Participants who were engaged in work at the interview locations clearly focused on the potential for the robot dog to assist them in their work tasks, e.g., cleaning or guiding customers to find books in the library. Interestingly, the noise made by the robot dog – emanating from the motors moving its legs and the impact of its feet on the floor – was only noted in the library environment. In this context, movement noise was the primary factor that caused participants to consider the use of robot dogs to be unsuitable.

Participants also identified general challenges with robot dogs working in public places, including technical malfunctions and the risk of fire. Concerns about vandalism, public fear of robots, and the robot's ability to recognize people were also raised. Additionally, the issue of the robot increasing electronic waste and the potential decline in genuine human interaction were noted.

The interviewees' responses to the RoSAS (Figure 13.5) generally aligned with the qualitative interview responses. The RoSAS includes three key dimensions: competence, warmth, and discomfort. The robot dog received a score of 63% in the competence dimension, suggesting that participants generally viewed the robot as capable, reliable, and knowledgeable. This high level of perceived competence indicates a strong degree of trust in the robot's ability to perform tasks effectively, which could be crucial for its acceptance in various social and functional contexts. The warmth dimension yielded a score of 50%. This score suggests that the robot is perceived as somewhat friendly, compassionate, and socially intelligent. While not as high as the competence score, this moderate level of perceived warmth implies that the robot is seen as more than just a utilitarian machine; it has the potential to engage in emotionally resonant interactions. Lastly, the robot received a score of 42% in the discomfort dimension. This low score indicates that the robot is generally not perceived as intimidating, scary, or unpleasant. While there may be some reservations, the robot dog does not appear to be a significant source of discomfort or fear among participants.

While the three subscales of the RoSAS generally exhibited high internal validity, the score for the word strange, which contributes to the discomfort subscale, stood out as an outlier. It was the second-highest-scoring attribute. We hypothesize that participants may have interpreted strange more in the context of novelty, attributing a more positive connotation to it than typically associated with discomfort. It's worth noting that other studies using the RoSAS have also reported higher ratings for strange compared to other items in the discomfort subscale (Jørgensen & Christiansen, 2021, Table 13.2). This suggests that the interpretation of this attribute may not be universally consistent.

Overall, these RoSAS scores provide valuable insights for designers and engineers, suggesting that the robot dog is well-positioned for social acceptance, given its high levels of perceived competence and relatively low levels of discomfort.

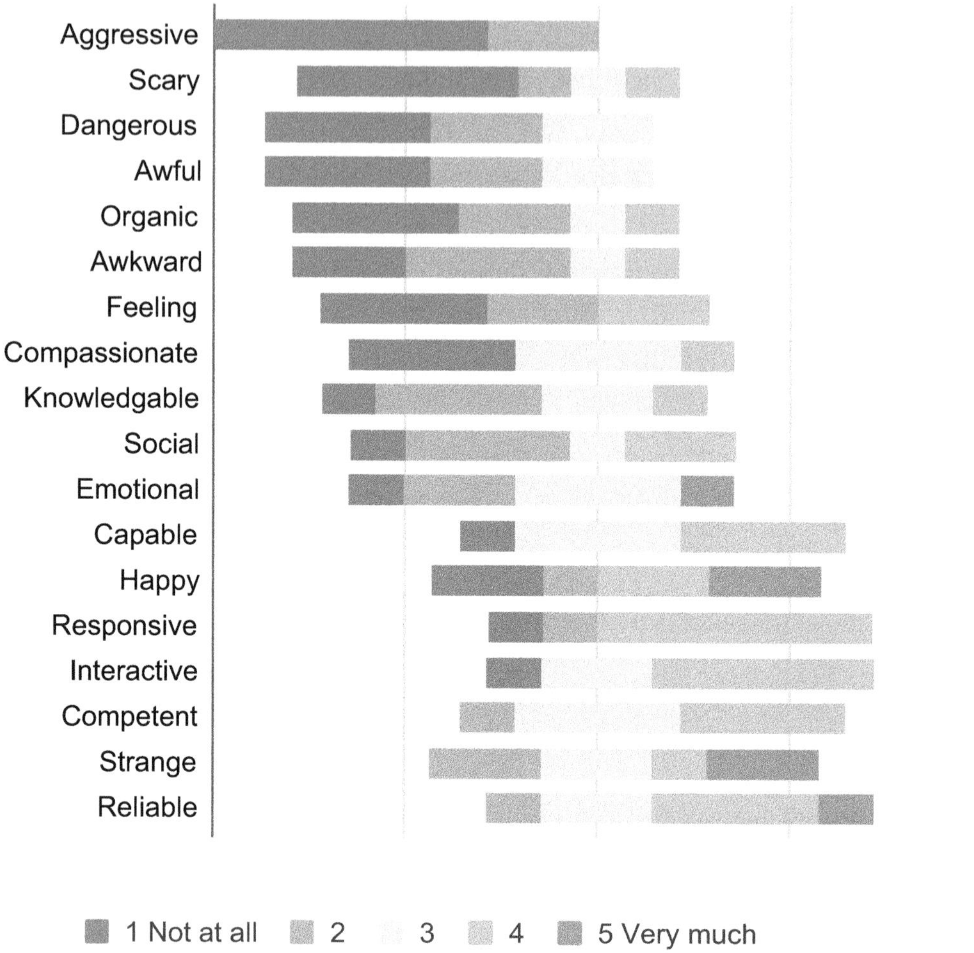

Figure 13.5 The Robotic Social Attribute Scale (RoSAS) (Carpinella et al., 2017) completed for the robot dog at the interview locations (N = 21).

Discussion

Based on our findings, we discuss issues related to the form of the robot dog and the importance of the context on acceptability, and present a model of the factors affecting the acceptance of robot dogs and the interconnections between them.

The Importance of Context

The location in which the robot dog is placed had a major influence on its acceptability. These location-based differences are relevant for smart cities, where robots may take on public-facing roles in transit, security, or maintenance. The airport environment emerged as a particularly favorable context for the robot dog; we speculate that there are two main reasons for this. First, the robot dog's aesthetic fits well with the airport's modern and technical ambiance, including a high level of ambient noise and activity. Second, the presence of dogs in airports for tasks like luggage inspection is already a familiar concept, paving the way for the presence of a robot dog.

Conversely, the cafe setting was considered less suitable for the robot dog. Participants felt that the intimate and social nature of cafes, where people engage in relaxed conversations, was at odds with the presence of a robot dog. Additionally, the limited physical space in cafes was cited as a constraint. The presence of a robot dog would undoubtedly create considerable disruption. This observation aligns with research by Cheon et al. (2022), who noted that the introduction of robots into various settings can precipitate complex social and organizational changes, affecting the dynamics between humans.

The Robot Dog Form Factor

A related finding from both studies was that participants consistently perceived the robot as a dog rather than a generic machine, and this form strongly influenced their expectations and reactions. This distinction influenced the types of tasks they envisioned for the robot dog, which often mirrored functions traditionally associated with dogs – such as scanning and searching – rather than tasks one might associate with humanoid robots. Similarly, participants did not expect the robot dog to speak to them, even though this would be a general expectation, e.g., based on the ubiquity of voice assistants such as Amazon Alexa. This specificity was either reflected in the tasks suggested or explicitly stated by participants. Additionally, the form factor influenced the robot's approachability. For instance, participants who had a fear of dogs found that this fear extended to the robot dog. This aligns with the findings of Paananen et al. (2023), who reported that in their study, participants often expressed a desire to interact with the robot dog in a manner similar to a real dog, such as petting or touching. Our study participants' focus on appearance over functionality echoes the findings of prior work that has noted the immediacy and strength of aesthetics on perceptions of HCI artifacts (Tractinsky, 2004).

Acceptance Model

Based on the findings from our two studies, Figure 13.6 presents the factors determining the acceptance of a specific robot dog application implementation. The Form Factor includes specific design elements such as the eyes, tail, and ears of the robot dog. These were missing from the robot dog model used in the studies, but were highlighted as critical for human interaction with what was perceived foremost as a dog. These elements not only influence Task Suitability

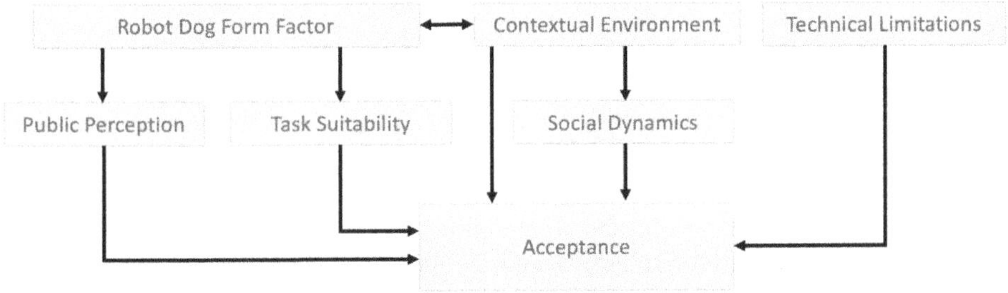

Figure 13.6 System diagram showing the factors affecting the acceptance of a robot dog performing tasks in a public setting, and the interconnections between them.

but also impact Public Perception, i.e., the broad, general attitude toward robot dogs, and Social Dynamics. The functionality of the tail and ears could serve as communication cues, as in real dogs, affecting both public perception and social interactions. As reported from our studies, there is a direct interaction between the visual appearance of the robot (form factor) and the Contextual Environment, with a match between the two contributing positively to acceptance. The Public Perception and Social Dynamics nodes are tightly linked, as the way the robot is perceived directly influences how it integrates into social settings, which in turn affects its acceptance. Task Suitability is influenced by the specific context in which the robot operates, such as an airport or a cafe, each having a unique set of requirements and limitations. A feedback loop exists between Public Perception and Acceptance, as the robot gains acceptance, e.g., if robot dogs become a common sight, its public image will likely improve, enhancing its acceptance.

Methodological Notes

Engaging citizens when assessing emerging technologies for smart cities is important, as this provides insight for technology acceptance and potential use cases. Combining different methodological approaches provides a more comprehensive overview to the possible challenges for technology adaptation. We speculate that in the studies utilizing image-based probes of robot dogs, this led participants to give visually oriented answers, e.g., how the robot dog's appearance (e.g., metallic materials) suited the particular context. Interestingly, these visually oriented comments were completely absent in the field interviews, where participants saw the actual robot dog in context. Rather, in the field interviews, comments related to factors such as the robot's functionality, the emotional appeal or the lack of it, and the quadrupedal form factor. This further differentiates our results from lab-based research, which may overlook experiential factors such as noise, motion awkwardness, or passers-by's reactions that influence acceptance in actual use contexts.

We wish to emphasize the importance of in-the-wild studies and physical probes when investigating citizen's perceptions on new technology solutions. We note that using a physical robot dog as the stimulus for evaluation is important in order to gain insight into holistic, multimodal experiences. For example, the noise of the robot dog's steps was particularly commented on – a factor that is hard to imagine based on images or through pictures or even a video.

We acknowledge that our research is limited by the cultural context the study was conducted in. Moreover, the field study participants' comments can be influenced by the novelty effect when encountering the robot dog for the first time. As a future work, a comparative cross-cultural study would be interesting to conduct.

Conclusion

Building on both survey and in-the-wild interviews, this paper explored how people perceive and interact with robot dogs across varied public settings. Utilizing a multi-method approach, we gathered rich data from field interviews and an online survey. Our field interviews offered immediate reactions and behavioral insights as participants interacted with a robot dog in real-world settings, such as countryside paths and shopping centers. The online survey provided a larger sample, highlighting general attitudes toward robot dogs.

Our findings highlight the importance context plays in human-robot interactions. For example, while robot dogs were generally well-received in public spaces like air- ports, their presence was questioned in more intimate settings like cafes. Moreover, the potential roles and tasks for robot dogs varied significantly depending on the environment, ranging from surveillance and security to emotional support. Potential challenges, such as technical breakdowns, vandalism, and societal concerns about the increasing role of robots in public spaces, were also raised by our study.

Acknowledgments

This work has been co-funded by the European Regional Development Fund (ERDF) and Lapin liitto as part of the AI.R and ProtoXR projects.

References

Abrams, A. M., Dautzenberg, P. S., Jakobowsky, C., Ladwig, S., & Rosenthal-Von Der Pütten, A. M. (2021). A theoretical and empirical reflection on technology acceptance models for autonomous delivery robots. In *Proceedings of the 2021 ACM/IEEE International Conference on Human-Robot Interaction* (pp. 272–280). ACM.

Breazeal, C. (2004a). *Designing sociable robots*. MIT Press.

Breazeal, C. (2004b). Social interactions in HRI: The robot view. *IEEE Transactions on Systems, Man, and Cybernetics, Part C (Applications and Reviews)*, *34* (2), 181–186.

Carpinella, C. M., Wyman, A. B., Perez, M. A., & Stroessner, S. J. (2017). The robotic social attributes scale (Rosas) development and validation. In *Proceedings of the 2017 ACM/IEEE International Conference on Human-Robot Interaction* (pp. 254–262).

Cheon, E., Schneiders, E., Diekjobst, K., & Skov, M. B. (2022). Robots as a place for social- izing: Influences of collaborative robots on social dynamics in-and outside the production cells. *Proceedings of the ACM on Human-Computer Interaction*, *6* (CSCW2), 1–26.

Chernyak, N., & Gary, H. E. (2019). Children's cognitive and behavioral reactions to an autonomous versus controlled social robot dog. In P. J. Marshall & K. Brenneman (Eds.), *Young children's developing understanding of the biological world* (pp. 73–90). Routledge.

Darling, K. (2016). Extending legal protection to social robots: The effects of anthropomor- phism, empathy, and violent behavior towards robotic objects. In Calo, R., Froomkin, A. M., & Kerr, I. (Eds.), *Robot law* (pp. 213–232). Edward Elgar Publishing.

Hegel, F., Lohse, M., & Wrede, B. (2009). Effects of visual appearance on the attribution of applications in social robotics. In *ROMAN 2009-the 18th IEEE international symposium on robot and human interactive communication* (pp. 64–71). IEEE.

Henschel, A., Laban, G., & Cross, E. S. (2021). What makes a robot social? A review of social robots from science fiction to a home or hospital near you. *Current Robotics Reports*, *2* (1), 9–19.

Jørgensen, J., & Christiansen, M. B. (2021). The sounds of softness. Designing sound for human-soft robot interaction. *Frontiers in Robotics and AI*, *8*, 674121.

Kahn Jr, P. H., Friedman, B., Perez-Granados, D. R., & Freier, N. G. (2004). Robotic pets in the lives of preschool children. In *CHI'04 extended abstracts on human factors in computing systems* (pp. 1449–1452). ACM.

Koelle, M., Ananthanarayan, S., & Boll, S. (2020). Social acceptability in HCI: A survey of methods, measures, and design strategies. In *Proceedings of the 2020 CHI Conference on Human Factors in Computing Systems* (pp. 1–19). ACM.

Leite, I., Martinho, C., & Paiva, A. (2013). Social robots for long-term interaction: A survey. *International Journal of Social Robotics*, *5*, 291–308.

Li, D., Rau, P. P., & Li, Y. (2010). A cross-cultural study: Effect of robot appearance and task. *International Journal of Social Robotics*, *2*, 175–186.

Löffler, D., Dörrenbächer, J., & Hassenzahl, M. (2020). The uncanny valley effect in zoomorphic robots: The U-shaped relation between animal likeness and likeability. In *Proceedings of the 2020 ACM/IEEE International Conference on Human-Robot Interaction* (pp. 261–270). ACM.

Martinez, J. E., VanLeeuwen, D., Stringam, B. B., & Fraune, M. R. (2023). Hey?! What did you think about that robot? Groups polarize users' acceptance and trust of food delivery robots. In *Proceedings of the 2023 ACM/IEEE International Conference* on *Human-Robot Interaction* (pp. 417–427). ACM.

McNelis, J., Schwall, G. J., & Collins, J. F. (2012). Robotic remote presence technology in the surgical intensive care unit. *Journal of Trauma and Acute Care Surgery*, *72*(2), 527–530.

Melson, G. F., Kahn Jr, P. H., Beck, A. M., Friedman, B., Roberts, T., & Garrett, E. (2005). Robots as dogs? Children's interactions with the robotic dog aibo and a live Australian shepherd. In *CHI'05 extended abstracts on human factors in computing systems* (pp. 1649–1652). ACM.

Millar, R., & Tracey, A. (2009). The interview approach. In Hargie, O. & Tourish, D. (Eds.) *Auditing organizational communication: A handbook of research, theory and practice*, 78–102. Routledge.

Morovitz, M., Mueller, M., & Scheutz, M. (2017). Animal-robot interaction: The role of human likeness on the success of dog-robot interactions. In *Proceedings of the 1st International Workshop on Vocal Interactivity in-and-between Humans, Animals and Robots (VIHAR), London, UK* (pp. 29–30).

Moshkina, L., Trickett, S., & Trafton, J. G. (2014). Social engagement in public places: A tale of one robot. In Proceedings of the 2014 *ACM/IEEE International Conference* on *Human-Robot Interaction* (pp. 382–389). ACM.

Neustaedter, C., Venolia, G., Procyk, J., & Hawkins, D. (2016). To beam or not to beam: A study of remote telepresence attendance at an academic conference. In Proceedings of the 19th *ACM Conference* on *Computer-Supported Cooperative Work* & *Social Computing* (pp. 418–431). ACM.

Niemelä, M., Arvola, A., & Aaltonen, I. (2017). Monitoring the acceptance of a social service robot in a shopping mall: First results. In Proceedings of the *Companion* of the 2017 *ACM/IEEE International Conference* on *Human-Robot Interaction* (pp. 225–226). ACM.

Paananen, S., Kalving, M., Colley, J., Kirjavainen, E., & Narbrough, M. (2023). *Quadrupedal robot dog as an attraction in a local outdoor event for live streaming and engagement.* Mensch und Computer 2023 – Workshopband. GI.

Sandry, E. (2015). Re-evaluating the form and communication of social robots: The benefits of collaborating with machinelike robots. *International Journal of Social Robotics*, *7*, 335–346.

Strauss, A. L. (1987). *Qualitative analysis for social scientists*. Cambridge University Press.

Sung, J., Christensen, H. I., & Grinter, R. E. (2009). Robots in the wild: Understanding long-term use. In Proceedings of the 4th *ACM/IEEE International Conference* on *Human Robot Interaction* (pp. 45–52). ACM.

Tractinsky, N. (2004). Toward the study of aesthetics in information technology. In *Proceedings of the International Conference on Information Systems (ICIS 2004), December 12–15, 2004,* Washington, DC, USA. Association for Information Systems.

Unitree. (n.d.). *Unitree robot go1 quadruped robot dog*. Retrieved 2023-09-13, from https://shop.unitree.com/

Väätäjä, H., Tapio, S., & Häkkilä, J. (2023). Exploring dogs' reactions when encountering delivery robots in urban environment. In *Proceedings of the 26th International Academic Mindtrek Conference* (pp. 350–353). ACM.

Walker, M., Hedayati, H., Lee, J., & Szafir, D. (2018). Communicating robot motion intent with augmented reality. In *Proceedings of the 2018 ACM/IEEE International Conference on Human-Robot Interaction (HRI '18)* (pp. 316–324). ACM. https://doi.org/10.1145/3171221.3171253

Weiss, A., Wurhofer, D., & Tscheligi, M. (2009). "I love this dog"—children's emotional attachment to the robotic dog aibo. *International Journal of Social Robotics*, *1*, 243–248.

Wu, C.-H., Ahtinen, A., & Väänänen, K. (2023). Walking outdoor with a zoomorphic mobile robot: Exploration of robot-assisted physical activities for older adults. In *2023 32nd IEEE International Conference on Robot and Human Interactive Communication (ro-man)* (pp. 269–276). IEEE.

Xiao, L., & Kumar, V. (2021). Robotics for customer service: A useful complement or an ultimate substitute*? Journal of Service Research*, *24* (1), 9–29.

Yadollahi, E., Couto, M., Dillenbourg, P., & Paiva, A. (2022). Motivating children to practice perspective-taking through playing games with cozmo. In *2022 31st IEEE International Conference* on *Robot* and *Human Interactive Communication* (ro-man) (pp. 1482–1489). IEEE.

14 Engaging Stakeholders in Prototyping with Immersive Theatre Practices and Worldbuilding on the Virtual Production Stage

Ella Björn, Satu Miettinen, Tuukka Uusitalo, Jyrki Kaukomies, Marjuska Kaukomies, Henna Huotarinen, and Reetta Kupari

Introduction

Rapid technological transformation guides organisations to adjust, imagine, and prototype near-future services. A more comprehensive and layered understanding of the future is needed (Slaughter, 1996), which can be achieved through diverse methodologies and perspectives. Although art-based methods have proven to foster creativity in business (Nissley, 2010), companies have not yet utilised them effectively for envisioning near-future changes in their operations. Participatory approaches applied in speculative design aim to enrich and engage participants in co-design practices (Farias et al., 2022) and foster reflection (Volkmann et al., 2023). Immersive theatre practices and worldbuilding derive from artistic disciplines such as performance, visual arts, and theatre, enhancing socially meaningful knowledge creation as well as embodied and affective engagement (Nissley, 2010; Rolling, 2013). This chapter explores how immersive theatre practices and worldbuilding can serve as innovative methods for engaging stakeholders, especially companies, in near-future service prototyping.

The experimental gulf between the abstract imagination of the future and its lived experience emphasises the need to construct artificial future spaces anchored in the embodied and affective dimensions of the present (Candy & Dunagan, 2016). Worldbuilding focuses on creating physical prototypes of imaginary future worlds, particularly in the literature, film, and gaming industries. It has also been utilised in the Research through Design (RtD) process to explore how creative methods and artistic expression can support companies' foresight and strategy work (Miettinen et al., 2025). This chapter examines how a physical filmmaking environment can be used in service prototyping.

Prototyping is a core element of design processes, helping to make abstract ideas more concrete, experiential, and tangible (Coughlan et al., 2007). Prototypes can range from low-fidelity sketches to physical artefacts and high-fidelity simulations, depending on the design's purpose (Bordegoni et al., 2023). Low-fidelity prototypes and speculative stories can serve as collaborative methods to represent future scenarios through potential technological applications and evolving roles within organisations (Aranda-Muñoz et al., 2025). In this research, prototyping is not limited to the creation of physical artefacts but also embraces experimental and performative approaches. To better understand the role of embodied engagement in near-future prototyping, the study co-produces and prototypes imaginary yet plausible, and in some cases, already emerging, near-future solutions through worldbuilding and immersive theatre. This approach aims to

DOI: 10.4324/9781003667872-18

unravel near-future trends that are already being experienced in the present. These practices are carried out on the virtual production stage.

The virtual production stage refers to a physical filmmaking environment equipped with computer-aided production tools, computer-generated imagery (CGI), LED screens, and game engine technologies (Li et al., 2022). An immersive space can be defined as an environment that activates mental processes occurring when one is psychologically drawn into a virtual world, focusing on experience and creating a sensation of presence within a narrative (Cummings & Bailenson, 2016). Immersive spaces encompass technologies such as Virtual Reality (VR), Augmented Reality (AR), and Extended Reality (XR). This chapter explores how the inclusion of diverse stakeholders and roles takes place through immersive theatre practices on a virtual production stage equipped with an LED wall, a game engine, and real-time rendering, where physical and virtual worlds converge. Although technologies like CGI, VR, and AR already exist and are in use, they are not yet commonly applied in near-future service development. There is a need for research that develops new practices that are accessible and applicable to companies. Service prototyping, as a participatory practice, enables companies to engage in this research and development work (Soto et al., 2023).

The chapter examines how bodily experience is realised in prototyping on a virtual production stage using CGI. The research provides empathic and embodied knowledge (Höök, 2018) on how to engage stakeholders in immersive spaces that emerge from encounters with sensorial, material, and technological elements (Pink et al., 2019) through imagining, performing, and prototyping. These encounters offer touchpoints to step into near-future worlds and focus on the sensations they evoke in our bodies.

The chapter presents a case study from the *Empathy Business* and the *Virtual Production Studio Technologies* projects, where stakeholders from design and research fields co-created and enacted near-future service scenarios for the mobility and automotive sectors. The process began with a workshop to develop near-future visions for mobility and automotive services, involving a car retail company, researchers, and designers. It continued with the prototyping of three selected near-future scenes on the virtual production stage: a car crash, a car service centre, and an AI home assistant. These scenes were prototyped using worldbuilding and immersive theatre practices to engage participants in envisioning and experiencing near-future service scenarios.

Through a combination of speculative design, immersive theatre, and worldbuilding, participants explored near-future mobility services in a virtual production environment. These scenarios were not only visualised but also performed, felt, and reflected upon, highlighting the potential of immersive prototyping to bridge the gap between abstract foresight and lived experience. Research data were collected through reflective discussions and a survey conducted with participants involved in the prototyping process. This approach emphasises the role of immersive technologies, bodily experience, emotion, and narrative in understanding near futures. The results showed that prototyping on the virtual production stage engaged participants in embodied co-creation and co-living of near-future scenarios, offering new insights into participatory design practices.

Engaging Stakeholders in Participatory Design

Participatory design involves users in multidisciplinary co-design processes, making technologies, tools, environments, businesses, and solutions more responsive to human needs (Simonsen & Robertson, 2013). Empathy is a critical element in these processes (Meroni & Sangiorgi, 2011), where user-centric and immersive approaches are employed to better understand real-world experiences. These approaches focus on the nature of problems, including sensorial, bodily, and emotional aspects, which are explored through interaction with people (Mattelmäki et al., 2014). Empathic co-design processes enable the identification, sharing, and integration of stakeholders' diverse worldviews, values, experiences, knowledge, perspectives, and ambitions (Smeenk, 2023). Active perspective-taking can be cultivated through techniques such as role-playing, gamification, and other immersive activities (Visser & van Erp, 2023). These methods not only enhance trust but also foster shared ownership in design processes.

Effective engagement highlights the role of emotions and embodiment in knowledge creation (Höök, 2018). Our embodied self reflects the understanding that emotions are not only mental but also physical experiences. Novel techniques can provide deeper insights into felt life (McCarthy & Wright, 2004). Immersive experiences enabled by emerging technologies (Cummings & Bailenson, 2016) can strengthen emotional connections and collaborative understanding in participatory design. Worldbuilding and immersive theatre practices offer new forms of sensory and emotionally rich embodied engagement, fostering shared understanding in participatory design processes.

Worldbuilding as a Tool for Prototyping

Worldbuilding is rooted in design fiction. The concept of design fiction originates from Bruce Sterling's writings on science fiction, bridging the fields of design, science fact, and science fiction (Bleecker, 2009). According to an interview with Bosch (2012, p. 1), Sterling defines design fiction as "the deliberate use of diegetic prototypes to suspend disbelief about change." Creating speculative yet plausible representations of the future can be used not only as a design technique (Bleecker, 2009) but also as a form of corporate propaganda (Blythe, 2014). Worldbuilding offers a structured approach to envisioning and prototyping future scenarios. When combined with participatory perspectives, speculative futures can help participants rehearse and bring near-mundane futures closer (Knutz et al., 2020). Prototyping on a cinematic virtual production stage is an artistic process that draws on immersive theatre practices (Machon, 2016), cinematic diegetic prototypes (Kirby, 2010), and worldbuilding.

Worldbuilding involves imagining and developing artefacts as prototypes that represent near futures (Duggan et al., 2017). These artefacts serve as material embodiments of possible future solutions, produced through disciplined imagination (Zimmerman & Forlizzi, 2008). Such representations can take various media forms, including narratives, short stories, films, physical objects, and more (Blythe, 2014; Coulton et al., 2017). According to Soren Rosenbak, a scholar of design fiction, diegetic prototypes are not intended to predict the future but to offer plausible services or products that "help to bring the future alive" (Sinclair, 2019, p. 1). He also notes that design fiction enables us not only to envision but also to critique and provoke the development trajectories of possible futures. It opens up space for dialogue around the broader societal effects and implications of future alternatives (Duggan et al., 2017).

Video prototypes are used to depict near-future service use cases, with participants acting as users and service providers interacting with speculative systems (Mackay & Beaudouin-Lafon, 2023). Filmmaking and diegetic prototypes are employed to create design fictions for strategic foresight (Miettinen et al., 2025), convey experiences and emotions (Gilardi et al., 2016), enable hybrid forms of participation (Uriu & Arima, 2025), and mediate wearable design concepts (Wang & Juhlin, 2017).

In this study, cinematic narratives and visual frameworks are used to prototype potential near futures involving technologies such as AI and other emerging innovations in the mobility and automotive industries. These scenarios are performed and filmed, allowing stakeholders to immerse themselves in envisioned near-future services and experiences. Virtual production environments enhance this process by merging physical and virtual elements. In this context, worldbuilding, narration, and immersive theatre practices are leveraged as novel methods for engaging stakeholders in prototyping processes.

Immersive Theatre Practices

The embodied experience of future worldbuilding and filmmaking relies on immersive theatre approaches. Immersive theatre encompasses visceral and participatory audience experiences, establishing a world composed of space, scenography, sound, and duration (Machon, 2016). It is closely linked to cinematic space, which refers to the space created on the cinema screen through the techniques and language of cinema (Kuhn & Westwell, 2012). Cinematic virtual reality (CVR) merges storytelling and technology, offering a distinctive way for audiences to engage with narratives in immersive settings (O Fearghail et al., 2024). Both immersive theatre and cinematic space aim to engage participants emotionally through sensory stimulation (Freitag et al., 2020; Rizopoulos et al., 2023). Technology-assisted environments, such as virtual production stages, enable the simulation and prototyping of service situations in immersive ways.

The concept of presence, a sense of "being there", is a desired element in mediated environments such as virtual spaces (Cummings & Bailenson, 2016). The term immersive is used to describe performance practices that are experiential, offering multisensory engagement and participation for audiences (Machon, 2016). Individuals can make sense of their actions by acting as spectators of their own interactions and attendants to their actions and reactions (Machon, 2016, p. 34). Immersive experience considers a participant's lived experience and physical body as responsive to imaginary and sensual environments (Machon, 2013, p. 61; Fällman, 2003). Immersive practices entail embodied watching, where the observer is actively engaged, not only by looking but also by feeling the moment (Machon, 2016).

When combined with design fiction methodologies such as diegetic prototypes and worldbuilding, immersive theatre offers new practices for participatory design processes that span both physical and virtual environments. Cinematic diegetic prototypes "allow scientists and filmmakers to visualise specific methods and technologies within the social realm of the fictional world" (Kirby, 2010, p. 66). These methodologies support the construction of collective memories through embodiment and presentness. Collective memories can be created through film by offering opportunities to archive and approach the future (Bačun, 2023). Experiential futures rely on "the continuum of human experience" (Candy & Dunagan, 2016, p. 137), and the embodied experience of filmmaking helps to address, live, and archive collective memories for the near future.

Research Design

The research process consisted of three phases: (1) defining future images during a collaborative workshop, (2) prototyping, performing, and filming future scenarios in a virtual studio, and (3) creating three videos based on these scenarios (see Figure 15.1). These videos served as the final output for the car company involved in the project. The workshop was attended by nine participants, including representatives from a car retail company, design researchers, and creative companies. The prototyping, performing, and filming sessions in the virtual production studio involved 15 participants: design researchers, filmmakers, virtual production technicians

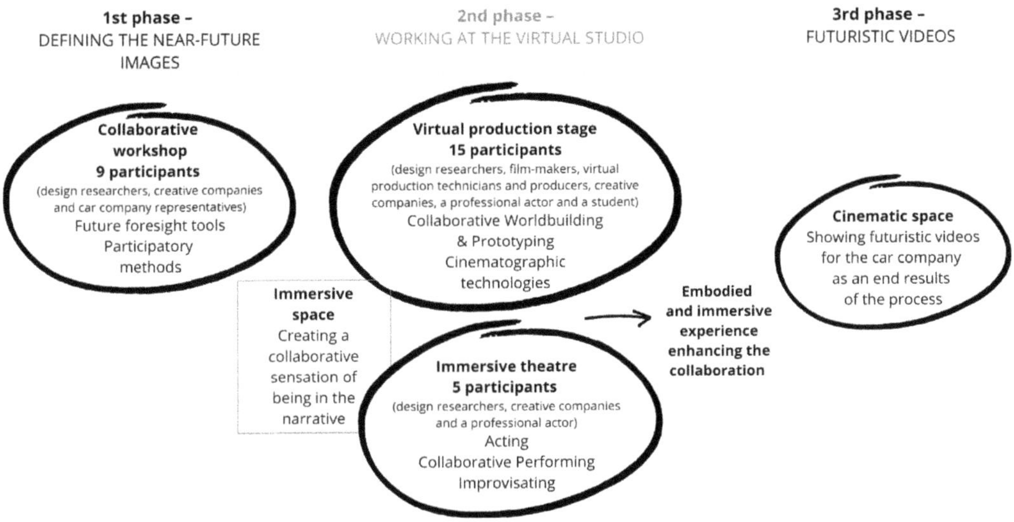

Figure 14.1 The process of engaging stakeholders in future envisioning through participatory and immersive approaches.

and producers, creative companies, a professional actor, and a student. Of these, five participants took part in acting and performing.

This research focused on a case study exploring the near future of mobility. The process included a workshop aimed at discussing and envisioning near-future scenarios and identifying trends that are already emerging within the mobility and automotive industries. These scenarios were designed to foster understanding of future users' emotions and experiences, and to explore how new technologies might enhance products and services in the car industry. The goal was not to produce commercial products, but to generate knowledge through design practice (Zimmerman & Forlizzi, 2008). The workshop resulted in five near-future concepts, from which three selected future images were enacted and filmed on a virtual production stage. This stage was in an audiovisual media lab and equipped with an LED wall and near-future prototypes, which were used to create three distinct scenes.

Data collection included note-taking and photography during reflective discussions in the workshop, as well as a survey conducted with participants involved in prototyping, worldbuilding, performing, and filmmaking on the virtual production stage. The survey was completed by six participants: two design researchers, one virtual production producer, one professional actor, and two representatives from creative companies.

Case Study: The Near-Future of Mobility – Foresight in the Automotive Sector

Since the future cannot be predicted, it is more appropriate to speak of different perceptions and images of the future (Nuutinen et al., 2017). The future of mobility is undergoing significant transformation, and the automotive sector must anticipate shifts in its operational environment. These changes are driven by factors such as environmental regulations, demographic trends, technological advancements (e.g., AI and autonomous vehicles), evolving societal values, and changing patterns in mobility behaviour.

1st Phase – Defining the Near-Future Images

The study began with a participatory workshop involving stakeholders from a car retail company, aimed at exploring anticipated developments in mobility and the automotive sector. Participants collaboratively examined emerging trends and co-created speculative scenarios, with a particular focus on future service models within car retail. This ideation process laid the foundation for selecting and prototyping plausible near-future concepts.

The workshop began with personal storytelling, allowing participants to connect with the themes on an individual level. Ecosystem mapping was then used to envision near-future changes and identify key actors within related service ecosystems. This method provided a structured yet flexible approach to exploring complex systems, relationships, and future possibilities. The service ecosystems perspective offers a more holistic understanding of value co-creation and insights into how different actors can influence it (Vink et al., 2021). An ecosystem map was developed to identify key stakeholders, their interactions, and external factors influencing future services within the automotive industry (see Figure 14.2.). The discussion also focused on the emotions and experiences of near-future users of car services. This process led to the creation of envisioned near-future scenarios for mobility and automotive services.

According to Rosenbak, while design fiction is based on speculative yet plausible representations of the future, it is grounded in researched knowledge (Sinclair, 2019). The near-future visions developed in this study were informed by research on current trends and anticipated revolutions in the mobility and automotive industries, as well as insights gathered during the workshop with a car retail company. Fulton (2018) identifies three major revolutions in urban transportation: vehicle electrification, automation, and shared (on-demand) mobility. The workshop discussions aligned with these trends and resulted in five near-future concepts:

1. Car as a service – Smart cars send information from the driver and vehicle to a call centre.
2. Short-term electric car rental – Flexible, temporary access to electric vehicles.

Figure 14.2 Ecosystem mapping for near-future car and mobility services.

3. Selling transportation, not cars – Mobility as a service rather than product ownership.
4. Modular cars – Vehicles with interchangeable components for different needs.
5. Social mobility – Cars as spaces for social connection and education, e.g., buses for elderly people and schools.

Based on the workshop discussions and outcomes, the concept of Car as a Service was selected for prototyping and performance in three distinct scenes on the virtual production stage. This was identified as a current trend in the automotive industry, driven by increasing demand for efficiency and comfort. Existing solutions already allow customers to avoid managing vehicle functionalities, as smart cars can send signals to a call centre, enabling remote diagnostics and partial issue resolution without the customer's awareness. Additional topics discussed included security concerns, driver emotions, and technological developments related to near-future driving.

2nd Phase – Working at the Virtual Production Stage

The Car as a Service near-future vision was approached from multiple perspectives. This concept was closely tied to the user experience of driving in the near future, which was depicted through three distinct scenes. The basic storylines for these scenes were collaboratively ideated prior to the filming session. Based on this ideation, three virtual environments were created using the Unreal Engine 5 game engine. These environments included:

- a wintry road,
- a futuristic service centre, and
- a living room featuring futuristic service possibilities.

These virtual settings served as backdrops for the creation of three future scenarios in the virtual production studio, establishing the necessary profilmic space for prototyping.

The mixed reality setup began with designing a storyline for each scene and selecting appropriate props (see Figures 14.3 and 14.4). Participants used their imagination to develop compelling scripts and choose props that would best convey the desired depth and atmosphere of each scene, drawing on the expertise of the filmmakers. For example, a futuristic robot dog was included in the living room scene to enrich the environment and support the near-future-oriented narrative.

Materials played a significant role in the scenario-building process at the virtual production stage. As Schön (1992, p. 133) describes, the design process is "a reflective conversation with the materials." Both physical and virtual materials were utilised in creating future images and filming environments. The worldbuilding process began by setting up the filming environment using physical materials such as broken windshields, car mirrors, and futuristic furniture. Lighting was carefully arranged to achieve the desired atmosphere for each scene (see Figure 14.5). These elements contributed to the immersive quality of virtual production and supported the narrative depth of the near-future scenarios.

The studio's LED wall was used to create immersive backdrops, while realistic physical sets were constructed within the studio space for each environment. This integration of physical and virtual elements enabled an immersive experience that allowed actors to fully engage with the envisioned future scenarios. Virtual elements were especially prominent in the AI home assistant scene, which featured a futuristic hologram projected on the wall (see Figure 14.6). The space was carefully designed to support the filming of each scenario, ensuring that the virtual and

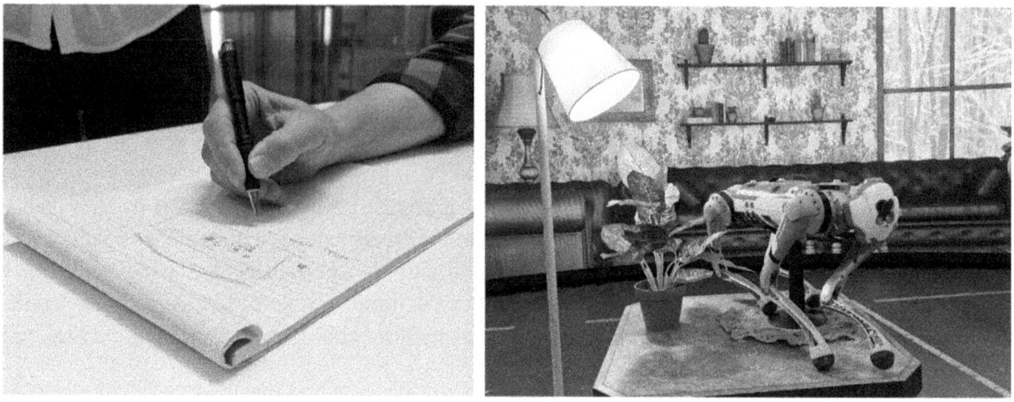

Figure 14.3 Designing the script storyline and selecting suitable props.

NAME OF THE SCENE	STORYLINE	PROPS USED
Car Crash	A car crash situation in which the driver actively communicates with the car. The AI assistant in the car assists the driver in choosing music and gives information about his current health status, such as stress level and the stability of vital functions after a crash, by calming down the driver.	• Wintery road with smoke in a wooded environment created with Unreal Engine 5 • Broken windshield • Gaming chair with steering wheel • Boxes and other supporting props • Suitable lightning
Car Service Center	The scene deals with a near-future service centre that receives and transmits situational information and analysis about traffic, weather conditions and other possible mobility-related happenings or threats in the area. In addition to AI assistants, the service centre also employs humans.	• Futuristic service centre environment created with Unreal Engine 5 • Tables and chairs • Charging pillars • Clothes and an energy drink • Suitable lightning
Home AI Assistant	The scene takes place at the user's home, where the user asks the AI assistant called Hilda to help plan a birthday trip for his wife and book a suitable car for the trip. The car is booked using a hologram projected onto the living room wall.	• Old living room environment with a hologram on the wall created with Unreal Engine 5 • Old furniture, plants and table cloths • Robotic dog • Suitable lightning

Figure 14.4 Three near-future scene descriptions and props used in worldbuilding.

physical worlds worked in conjunction to create viable yet imaginative representations of the near future. By building the environments as virtual productions, real-time transitions between scenes were made possible, allowing actors to remain fully immersed in the experience and enhancing the authenticity of the prototyping process.

Immersive theatre, filmmaking, and worldbuilding practices were employed to foster embodied, sensorial engagement between lived experience and the imagined near future of the driver, the car, and the surrounding environment. Although the scenes and scripts were loosely planned by participants prior to filming, the filming session allowed room for improvisation. The script provided direction and structure, guiding participants toward the desired outcomes, while still allowing them to improvise and shape the scenes themselves.

Figure 14.5 Setting up the car crash (left) and service centre (right) scenes.

Figure 14.6 Performing and interacting with the physical and virtual materials in the *AI home assistant* scene.

3rd Phase – Futuristic Videos

The final phase of the process involved presenting the near-future visions in video format to representatives of the car retail company. These videos served as the culminating output, designed to support the company's foresight initiatives and facilitate discussions around anticipated developments in mobility. Feedback from the company highlighted the value of using such methods to effectively communicate near-future scenarios to potential collaborators and funders.

Findings and Discussion

Six participants completed the survey, providing feedback on their experiences working at the virtual production stage. The survey responses included input from design researchers, a virtual production producer, an actor, and representatives from a creative company. In addition to the written feedback, a reflective discussion session was held with seven participants, focusing on the use of immersive theatre practices and worldbuilding within the virtual production stage as a collaborative prototyping method. The discussion participants included service designers, filmmakers, one actor, and researchers.

Although most participants had a background in design and prior experience with prototyping, they had limited exposure to design fiction, performance, or immersive theatre practices. One participant was a professional actor with extensive experience in performance. The filmmakers had more expertise in virtual production technologies and worldbuilding than the other participants. While they did not respond to the survey, they actively contributed to the discussion session.

Worldbuilding and Prototyping on the Virtual Production Stage

According to the survey results, five out of six participants found worldbuilding and prototyping to be relatively easy, while one participant left the question unanswered.

> It was interesting to try out how different and differently sized Unreal scenarios work in filming situations.
>
> (P1)

> The workshop sparked a lot of ideas in terms of future anticipation and service design. Working with design fiction early in the design process may be the most important motivating and illustrative factor in terms of designing the desired future. This could be implemented for a wide variety of large organisations.
>
> (P2)

> The work reinforced the importance of using narrative methods and the significance of prototyping as a means of increasing understanding.
>
> (P3)

Prototyping on the virtual production stage demonstrated that the ability to quickly generate and edit background images using Unreal Engine was essential to the process.

> Utilising artificial intelligence in the future will speed up the creation of future images" and "utilising different data on the fly in connection with the creation of future images could be interesting.
>
> (P1)

> Future visions became visible, and it was easier to communicate and take them forward.
>
> (P4)

> I think what characterised this workshop was that the tools were tried out and their possibilities were explored.
>
> (P3)

Both the reflective discussion session and the survey results indicated that the working method was engaging and effective in illustrating possible future scenarios. However, one participant pointed out:

> To some extent, the high level of ambition in AV (audiovisual media) production meant that instead of quickly visualising ideas, attention and time went into optimising the AV end product, rather than quickly testing it.
>
> (P4)

Another participant emphasised the importance of ideation:

> I was interested in working in a way that focused a little more on the ideation of the content itself and a little less on the technical implementation, which would potentially allow for the creation of multiple future images at the same time—perhaps conflicting possible worlds—but with sufficiently good technical quality.
>
> (P2)

The reflective discussion also highlighted that setting up the stage requires time, but it is an integral part of the process that helps participants mentally engage with the envisioned future. The hands-on co-creation approach enabled participants to generate new ideas creatively and rapidly through ongoing interaction with both virtual and physical materials, what Schön (1992, p.133) describes as "a reflective conversation with the materials." Despite limited resources, the scenarios were perceived as plausible, underscoring the importance of collaborative work.

Embodied Experience and Performing on a Virtual Production Stage

According to the survey results, three out of six participants found performing relatively easy, while two felt neutral about the experience. Novel techniques can help embody the future (McCarthy & Wright, 2004), making it tangible in the present (Candy & Dunagan, 2016). One participant noted:

> Embodying helps to imagine. Viewers can also better imagine different futures when participants empathise and when artefacts are used. It is good to have at least one professional actor who can empathise and direct others.
>
> (P2)

Another participant emphasised the importance of interaction:

> Experiencing interaction is a very important part of understanding.
>
> (P4)

One participant (P5) described the experience as positive and refreshingly different from their usual work, though they noted that the short duration of the scene might have benefited from a warm-up to enhance engagement. Another participant (P3) acknowledged the typical challenges of performing, such as overcoming self-consciousness, but highlighted that the calm and professional atmosphere of the production team helped ease the process.

Reflective discussions revealed that the roles of the characters evolved throughout the filming days, which ultimately enhanced the actors' performances. This suggests that acting and performing became easier as the group grew more familiar with working together. Participants also emphasised the importance of providing stimuli that promotes immersion and imagination, such as sound, lighting, and set design.

Many of the lines were improvised, particularly in scenes where final lines were not rehearsed in advance. This spontaneity added a genuine sense of empathy and authenticity to the performances, creating moments of what could be described as "art by accident" (see e.g., Hilde, 2000).

Immersive Theatre Practices and Worldbuilding as Tools to Build Engagement

Three out of six participants felt that these methods were quite useful for empathising with the future, while the remaining three believed they were helpful to some extent. One participant (P6) stated that performing enables one to imagine themselves in the character's shoes. However, one participant (P5) noted that the effectiveness of this approach depends significantly on the scene and role, particularly if the activity involves simultaneous acting and dialogue. Another participant highlighted the importance of co-design:

> The method emphasises the importance of co-design, where technological expertise was highlighted.
>
> (P1)

These reflections suggest that while immersive methods are generally effective, their impact can be influenced by the structure of the activity and the participants' familiarity with performative techniques.

According to both the reflective discussions and survey results, these methods were found to enhance shared understanding by building collective memories (Bačun, 2023). One participant remarked:

> Foresight, building future images, and creating scenarios is about creating stories together. If stories remained tables and maps, achieving a common understanding to work towards a common goal would be quite limited.
>
> (P4)

Another participant noted:

> Experiencing things together helps you understand the subject better and also creates a shared picture that helps with processing matters and builds empathy.
>
> (P3)

It was also observed that working in a closed environment over two intensive days fostered connections among participants and helped the team align toward a shared goal, while also having fun. Overall, the findings indicate that immersive and performative approaches not only support emotional resonance but also facilitate deeper interpersonal understanding and collective envisioning in participatory design contexts.

Conclusion

This chapter has explored how immersive theatre practices and worldbuilding, when combined with virtual production technologies, can serve as powerful tools for engaging stakeholders in prototyping near-future services. By situating participants within speculative yet plausible scenarios, this approach enabled embodied and empathic engagement with future possibilities, transforming abstract near-future concepts into lived, shared experiences that are more easily communicated and transferred within organisations.

The case study on the near future of mobility demonstrated how co-creating and performing near-future service scenarios on a virtual production stage fosters deeper understanding, emotional resonance, and collaborative ideation. Participants were not merely observers but active agents in shaping and experiencing future narratives. This embodied participation helped bridge the gap between imagination and implementation, suggesting that immersive prototyping environments can enhance stakeholder engagement, support interdisciplinary collaboration, and generate rich, multisensory insights into near-future services. While technical complexity and resource demands remain challenges, the creative and experiential value of these methods presents a compelling case for their broader adoption in strategic foresight and participatory design processes.

Ultimately, this research contributes to the evolving discourse on embodied engagement in design, highlighting the potential of immersive, performative, and narrative-driven approaches to prototype near futures that are not only seen but felt, shared, and co-created. It reframes the role of design fiction from a speculative storytelling tool to a lived, co-performed design practice, positioning immersive theatre and virtual production as critical infrastructures for participatory futuring.

Acknowledgements

This chapter is based on two funded projects. The Empathy Business project (duration 1.2.2023–31.12.2024) was funded by Business Finland and the Virtual Production Studio Technologies (VPST) (duration 1.08.2023–31.12.2025) is funded by the European Regional Development Fund (ERDF). Acknowledgements for Samuli Valkama supporting the process.

References

Aranda-Muñoz, Á., Yams, N. B., & Carlgren, L. (2025). Co-experiential futuring: Where speculative design and arts meet futures studies. *Futures*, *166*, 103549.

Bačun, N. (2023). Cinematic space in-between reality and fiction. In Holmlid, S., Rodrigues, V., Westin, C., Krogh, P. G., Mäkelä, M., Svanaes, D., & Wikberg-Nilsson, Å (Eds.), *Proceedings of Nordes 2023: This Space Intentionally Left Blank*, 12–14 June, Linköping University, Norrköping, Sweden. https://doi.org/10.21606/nordes.2023.131

Bleecker, J. (2009). *Design Fiction: A Short Essay on Design, Science, Fact and Fiction.* Near Future Laboratory.

Blythe, M. (2014). Research through design fiction: Narrative in real and imaginary abstracts. In Proceedings of the SIGCHI Conference on Human Factors in Computing Systems (CHI '14) (pp. 703–712). Association for Computing Machinery. https://doi.org/10.1145/2556288.2557098.

Bordegoni, M., Carulli, M., & Spadoni, E. (2023). Prototyping: Practices and Techniques. In *Prototyping User eXperience in eXtended Reality* (pp. 29–47). SpringerBriefs in Applied Sciences and Technology. Springer Nature Switzerland. https://doi.org/10.1007/978-3-031-39683-0_3

Bosch, T. (2012). *Sci-Fi Writer Bruce Sterling Explains the Intriguing New Concept of Design Fiction.* Slate. https://slate.com/technology/2012/03/bruce-sterling-on-design-fictions.html
Candy, S., & Dunagan, J. (2016). Designing an experiential scenario: The People who vanished. *Futures, 86*, 136–153. https://doi.org/10.1016/j.futures.2016.05.00
Coughlan, P., Suri, J. F., & Canales, K. (2007). Prototypes as (design) tools for behavioral and organizational change: A design-based approach to help organizations change work behaviors. *The Journal of Applied Behavioral Science*, *43*(1), 122–134. https://doi.org/10.1177/0021886306297722
Coulton, P., Lindley, J., Sturdee, M., & Stead, M. (2017). Design fiction as world building. In *Proceedings of the 3nd Biennial Research Through Design Conference* (pp. 1–16). RTD Conference.
Cummings, J. J., & Bailenson, J. N. (2016). How immersive is enough? A meta-analysis of the effect of immersive technology on user presence. *Media Psychology*, *19*(2), 272–309.
Duggan, J. R., Lindley, J., & McNicol, S. (2017). Near future school: World building beyond a neoliberal present with participatory design fictions. *Futures*, *94*, 15–23.
Fällman, D. (2003). *In Romance with the Materials of Mobile Interaction: A Phenomenological Approach to the Design of Mobile Information Technology*. Doctoral Thesis, Umeå University, Larsson & Co:s Tryckeri.
Farias, P. G., Bendor, R., & Van Eekelen, B. F. (2022). Social dreaming together: A critical exploration of participatory speculative design. In *Proceedings of the Participatory Design Conference 2022* (pp. 147–154), 2. ACM.
Freitag, F., Molter, C., Mücke, L-K., Rapp, H., Schlarb, D.B, Sommerlad, E., Spahr, C., & Zerhoch, D. (2020). Immersivity: An interdisciplinary approach to spaces of immersion. *Ambiances. Environnement sensible, architecture* et *espace urbain.* Varia. Online since 11 December 2020. http://journals.openedition.org/ambiances/3233
Fulton, L. M. (2018). Three revolutions in urban passenger travel. *Joule*, *2*(4), 575–578.
Gilardi, M., Holroyd, P., Brownbridge, C., Watten, P. L., & Obrist, M. (2016). Design fiction filmmaking: A pipeline for communicating experiences. In *Proceedings of the 2016 CHI Conference Extended Abstracts on Human Factors in Computing Systems* (pp. 1398–1406). ACM.
Hilde, T. C. (2000). Intelligence, accident, and art as a practice. *Transactions of the Charles S. Peirce Society*, *36*(4), 549–561.
Höök, K. 2018. *Designing with the Body: Somaesthetic Interaction Design*. The MIT Press.
Kirby, D. (2010). The future is now: Diegetic prototypes and the role of popular films in generating real-world technological development. *Social Studies of Science*, *40*(1), 41–70.
Knutz, E., Markussen, T., & aus der Wieschen, M. V. (2020). Probing the future of participatory healthcare through speculative design. In *Proceedings of the 6th European Design4Health Conference* (pp.73–82), 2. Lab4Living, Sheffield Hallam University.
Kuhn, A., & Westwell, G. (2012). *A Dictionary of Film Studies*. OUP Oxford.
Li, H., Lo, C.-H., Smith, A., & Yu, Z. (2022). The development of virtual production in film industry in the past decade. In *Proceedings of the 2022 International Conference on Innovation and Intelligence for Informatics, Computing, and Technologies.* Lecture Notes in Computer Science, vol 13312, (pp. 221–239), Springer, Cham.
Machon, J. (Ed.). (2013). *Immersive theatres: Intimacy and Immediacy in Contemporary Performance.* Palgrave Macmillan.
Machon, J. (2016). Watching, attending, sense-making: Spectatorship in immersive theatres. *Journal of Contemporary Drama in English*, *4*(1), 34–48.
Mackay, W. E., & Beaudouin-Lafon, M. (2023). Participatory design and prototyping. In Vanderdonckt, J., Palanque, P., & Winckler, M. (Eds.), *Handbook of Human Computer Interaction* (pp. 1–33). Springer International Publishing.
Mattelmäki, T., Vaajakallio, K., & Koskinen, I. (2014). What happened to empathic design? *Design Issues*, *30*(1), 67–77. https://doi.org/10.1162/desi_a_00249
McCarthy, J., & Wright, P. (2004). Technology as experience. *Interactions*, *11*(5), 42–43.
Meroni, A., & Sangiorgi, D. (2011). Introduction. In Meroni A., & Sangiorgi D. (Eds.), *Design for Services* (pp. 33–60). Routledge.

Miettinen, S., Sarantou, M., Björn, E., Uusitalo, T., & Kontio, T. (2025). Strategic foresight and worldbuilding with companies. In L. Imbesi, & A. Perlatti (Eds.), *Proceedings of the Cumulus 2024 Conference: The Design Across Borders: United in Creativity* (pp. 282–302). Cumulus.

Nissley, N. (2010). Arts-based learning at work: Economic downturns, innovation upturns, and the eminent practicality of arts in business. *Journal of Business Strategy*, *31*(4), 8–20.

Nuutinen, A., Räisänen, R., & Fernström, P. (2017). Ugly as a concept in craft to examine alternative futures. *European Journal of Futures Research*, *5*, 1–10.

O Fearghail, C., Gadipudi, N., & Young, G. W. (2024, June). Back to the virtual future: Presence in cinematic virtual reality. In *Proceedings of the 2024 ACM International Conference on Interactive Media Experiences Workshops* (pp. 80–87). ACM.

Pink, S., Fors, V., & Glöss, M. (2019). Automated futures and the mobile present: In-car video ethnographies. *Ethnography*, *20*(1), 88–107.

Rizopoulos, C., Arsenopoulou, N. & Poupou, A. (2023). Journeys, maps, dungeons: Navigating narrative and cinematic space in open world digital games. In *Proceedings of the 2nd International Conference of the ACM Greek SIGCHI Chapter* (pp. 1–5). ACM.

Rolling, J. H. (2013). Arts-based research: Systems and strategies. *Art Education*, *69*(3), 4–5.

Sarantou, M., & Miettinen, S. (2022). Introduction. In Sarantou M., & Miettinen S. (Eds.), *Empathy and Business Transformation* (pp. 1–12). Taylor & Francis.

Schön, D. A. (1992). Designing as reflective conversation with the materials of a design situation. *Research in Engineering Design*, *3*(3), 131–147.

Simonsen, J., & Robertson, T. (Eds.). (2013). *Routledge International Handbook of Participatory Design* (Vol. 711). Routledge.

Sinclair, J. (2019). *Design Fiction Portrays Possible Futures through Storytelling*. Aalto University Executive Education Oy, December 20. www.aaltoee.fi/en/aalto-leaders-insight/2019/design-fiction-portrays-possible-futures-through-storytelling

Slaughter, R. A. (1996). Futures studies: From individual to social capacity. *Futures*, *28*(8), 751–762.

Smeenk, W. (2023). The empathic co-design canvas: A tool for supporting multi-stakeholder co-design processes. *International Journal of Design*, *17*(2), 81–98. https://doi.org/10.57698/v17i2.05

Soto, M., Alhonsuo, M., & Miettinen, S. (2023). The prototyping service experience as a tool of empathy and research: From Finland to Chile. In Cipolla, C., Mont'Alvão, C., Farias, L., & Quaresma, M. (Eds.), *Proceedings of ServDes.2023 Entanglements & Flows Conference: Service Encounters and Meanings*, 11–14th July 2023, Rio de Janeiro, Brazil. Linköping Electronic Conference Proceedings 203. https://doi.org/10.3384/ecp203

Uriu, D., & Arima, S. (2025). Designing virtual funerals as a design fiction: A film-based exploration of near-future memorial rituals. In *Proceedings of the 2025 CHI Conference on Human Factors in Computing Systems* (pp. 1–19). ACM.

Vink, J., Koskela-Huotari, K., Tronvoll, B., Edvardsson, B., & Wetter-Edman, K. (2021). Service ecosystem design: Propositions, process model, and future research agenda. *Journal of Service Research*, *24*(2), 168–186.

Visser, F. S., & van Erp, J. (2023). Empathy building through virtual reality filmmaking in social innovation: Decreasing tensions between socially opposed citizens. *Strategic Design Research Journal*, *16*(1), 7–20.

Volkmann, T., Schaller, M., Nolte, A., Jochems, N., & Fischer, F. (2023). Deconstruction and reflection in participatory research and design–A multidisciplinary perspective. In *Mensch und Computer 2023-Workshopband* (pp. 10–18420). GI.

Wang, J., & Juhlin, O. (2017). Fashion, fiction, function: mediating wearable design through fashion film. In Stuedahl, D., & Morrison, A. (Eds.), *Proceedings of Nordes 2017: Design + Power*, 15–17 June, Oslo School of Architecture and Design, Norway. https://doi.org/10.21606/nordes.2017.016

Zimmerman, J., & Forlizzi, J. (2008). The role of design artifacts in design theory construction. *Artifact: Journal of Design Practice*, *2*(1), 41–45.

15 A Framework of Engaging Smart Technologies for Citizen Participation in the Lifecycle of Sustainable Built Urban Environments

Kaisa Väänänen, Nur Samancioglu, Jouko Makkonen, Raúl Castaño-Rosa, Anniina Autero, and Sofie Pelsmakers

Introduction

Urban environments face a complex challenge concerning environmental, economic, and social sustainability (Næss, 2001). Because cities are major contributors to climate change and resource depletion, they require planning that advances energy efficiency and reduced ecological impact (Bibri, 2020). Moreover, the well-being of citizens is crucial and involves not only access to healthy environments without pollution but also accessible spaces and fulfilling social interactions (Dempsey et al., 2011).

Involving citizens (i.e., all urban human stakeholders, i.e., residents, commuters, business owners, and visitors) is crucial in planning processes. A holistic, lifecycle-span approach is essential when integrating *citizen engagement* across different urban scales and supporting long-term sustainability with sustained cognitive, emotional, and behavioural interaction (Fredricks et al., 2004). This is often not achieved with current common digital platforms for citizen participation, as they are often limited to short-term interactions (AlWaer & Cooper, 2020), leading to "tokenism" – superficial inclusion without real influence – and missing out on local knowledge or meeting local needs (Chado et al., 2017).

However, emerging smart technologies offer promising avenues for deeper citizen engagement, e.g., 3D simulations, digital twins, GIS, BIM, IoT sensors, AI, and generative AI could be utilised in systems that support ongoing citizen involvement. These technologies can help create immersive experiences that support a sense of ownership and encourage real-time feedback (Berigüete et al., 2024) and transform citizen roles from passive participants to active co-creators, enabling their continuous input in urban development.

Current approaches for citizen participation are fragmented and short-term, limiting sustained engagement across different scales and phases of urban development. There is a need for a systematic approach for designing tools for continuous citizen engagement. Based on existing literature and insights from our own case projects, in this chapter we propose a novel, multi-dimensional framework called **UrbanEngagingTech** to facilitate designing ways for citizen engagement with urban environments based on smart technologies. This framework aims to be holistic, covering various scales – buildings, neighbourhood blocks, and cities – and different urban development lifecycle phases, including planning, construction, operation, and renewal. By integrating social and environmental sustainability perspectives, the framework provides a structured approach to re-think emerging smart technologies for participatory urban design.

DOI: 10.4324/9781003667872-19

The remainder of this chapter is structured as follows. Section "Background" reviews technological approaches that address environmental and social sustainability in urban planning, alongside attributes of engaging technology interactions for citizen engagement. Section "The Conceptual Framework for Technology-Mediated Citizen Participation: UrbanEngagingTech" presents the proposed UrbanEngagingTech framework in detail, integrating these perspectives across urban scales and development lifecycle phases. Section "Mapping the Case Projects and Engaging Technologies to the UrbanEngagingTech Framework" presents three case projects that illustrate the framework in practice: the INPERSO project (building scale), the Nordic Superblocks as Decarbonization Catalysts project (block scale), and the WeGenerate project (city/district scale). Section "Discussion and Conclusion" discusses the framework's implications, identifies gaps in current practice, and outlines considerations for equitable, long-term citizen participation.

Background

Technological Approaches for Environmental and Social Sustainability in Urban Planning

The growing emphasis on smart cities has often focused on economic growth while neglecting the balance required for environmental and social sustainability (Yigitcanlar et al., 2019). Bridging this gap requires the integration of smart technologies with sustainability principles to ensure inclusive, resilient, and ecologically responsible urban development (Bibri et al., 2023).

Smart technologies offer novel opportunities to optimise resource efficiency at both urban and building scales. Environmental sustainability can be advanced through smart grids and energy-efficient systems that optimise energy consumption, supported by IoT-based monitoring and control of HVAC, lighting, and other building services (Khan et al., 2017). At the urban scale, digital twins integrate environmental data to support green infrastructure planning and climate-resilient design (Gholami et al., 2024).

Smart technologies can facilitate social sustainability by improving participation, inclusivity, and well-being. Digital platforms and applications enable diverse communities to engage in urban decision-making, ensuring that the needs of different societal groups are reflected in policy and design processes (Wilson et al., 2019). At the building level, smart building systems enhance comfort and health via real-time environmental monitoring (Castaño-Rosa et al., 2024). However, poorly designed interfaces or oversimplified tools may risk distorting public input or excluding less technology-oriented groups, leading to incomplete representation of citizen perspectives (Berigüete et al., 2024).

Participatory urban planning increasingly leverages technology to overcome traditional barriers such as logistical constraints, unequal representation, and limited synthesis of citizen input (Fung, 2015). Immersive technologies such as Virtual Reality (VR) enable citizens to experience proposed developments in simulated environments, improving comprehension of spatial designs and meaningful co-creation processes (Jamei et al., 2017). Augmented Reality (AR) overlays proposed developments onto real-world contexts, which is particularly effective in adaptive reuse scenarios (Fanani et al., 2021). Participatory mapping tools, such as Public Participation Geographic Information Systems (PPGIS), allow stakeholders to contribute spatial knowledge, integrating local insights into urban planning models in an actionable manner (Fagerholm et al, 2021).

Artificial intelligence (AI) is emerging as a powerful enabler of participatory urban planning, extending to domains like mobility, energy management, and public health (Yigitcanlar et al., 2020). AI techniques also support analysis of unstructured public feedback, sentiment analysis, automated visualisation, and chatbots to facilitate communication between citizens and authorities (Foroughi et al., 2025). To achieve effective AI deployment in urban planning requires a human-in-the-loop approach, where AI-generated insights complement and support, rather than replace, deliberative democratic processes (Hadiyana & Ji-hoon, 2024).

Despite their transformative potential, smart technologies also introduce environmental and social risks that must be carefully managed. Increased energy consumption, material demands, and carbon emissions linked to the deployment of AI, IoT, and data centres could undermine environmental objectives (Sharma et al., 2023). Socially, digital divides, data privacy concerns, and potential reinforcement of existing inequalities demand critical attention to avoid exacerbating marginalisation and ensure equitable access and representation (Bibri et al., 2023). Thus, achieving genuinely sustainable urban development requires balancing the opportunities offered by technology with their inherent risks through careful design, governance, and continuous reflection.

Attributes of Engaging Technology Interactions for Citizen Engagement

Psychological engagement involves *cognitive engagement*, including attention, problem-solving, and mental effort; *emotional engagement*, including affective connection, enjoyment, and motivation, and *behavioural engagement*, including active participation and interaction (Fredricks et al., 2004). In Human-Computer Interaction, engagement is influenced by several related human factors principles that shape user interactions. *Trust, usability*, and *reliability* are basic design factors that impact sustained engagement with technology (Kaewkitipong et al., 2022). *Positive user experience* connects to personal attachment and emotional resonance (McCarthy & Wright, 2005). *Societal and ethical considerations* such as accessibility, privacy, and social responsibility are essential for responsible technology design (Hochheiser & Lazar, 2007).

Emerging HCI paradigms leverage *interactivity, immersion, and pervasiveness*, with multisensory technologies and VR seamlessly integrating into activities of daily life (Prestopnik & Zhang, 2015). Technology engagement builds on *challenge* (i.e., providing the right level of difficulty to sustain engagement), *positive affect (*i.e., ensuring enjoyment and satisfaction), and *aesthetic and sensory appeal* (i.e., making the interface appealing through human senses) (O'Brien & Toms, 2008). Furthermore, *persuasive design techniques* can be applied to drive behaviour change through human-technology interaction. Such techniques include *social features* to leverage social influence to persuade people, e.g., through social facilitation, cooperation or competition (Oinas-Kukkonen & Harjumaa, 2009), and *gamification* techniques such as challenges, comparison, and storytelling (Rapp et al., 2019).

In the context of urban planning, Fredericks et al. (2018) have explored how participatory technologies can enhance citizens reach and engage in city-making processes. They highlight the effectiveness of combining digital and analogue engagement channels to reach a broader demographic. Fredericks et al. (2020) outline *eight design patterns to guide smart engagement: raising awareness, creating interest, understanding local contexts, facilitating community discussions, and providing immediate feedback.* They also highlight the use of existing urban infrastructure to engage and connect diverse community members effectively.

The Conceptual Framework for Technology-Mediated Citizen Participation: UrbanEngagingTech

Citizen participation often occurs only in isolated development phases, producing reactive rather than proactive citizen input that addresses immediate concerns but may fail addressing

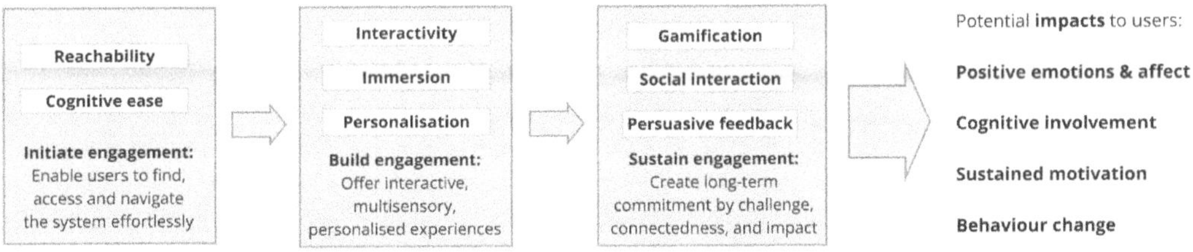

Figure 15.1 Design attributes of engaging smart technologies that can drive citizen engagement in urban development projects from the initial encounters to the sustained engagement.

long-term urban sustainability goals. We propose **UrbanEngagingTech** framework as a multi-dimensional approach to guide the design and integration of engaging smart technology solutions into participatory urban development across various scales and lifecycle phases. By intersecting urban scales (building, block, city) with construction lifecycle phases (planning, construction, operation, renewal) and sustainability perspectives (social and environmental) and attributes of engaging technologies for citizen engagement, this framework aims to provide a systematic model for embedding citizen participation throughout urban development.

Engaging technologies can foster participation and collaboration among diverse stakeholders. Based on the factors and techniques for technology-mediated citizen interactions identified in Section "Attributes of Engaging Technology Interactions for Citizen Engagement" we summarised the following eight **attributes of engaging technologies for citizen engagement**:

1. **Reachability:** Proactivity, accessibility, hybrid digital-physical interactions.
2. **Cognitive ease:** Reducing unnecessary complexity, usability, intuitive interactions.
3. **Interactivity:** Hands-on manipulation of objects and designs, real-time feedback, user control.
4. **Immersion:** Visualisations, multisensory interaction, aesthetics, real-world modelling.
5. **Personalisation:** Adapting to the user preferences and actions.
6. **Gamification:** Challenge, competition, achievements, rewards.
7. **Social interaction:** Connectedness with community, collaboration.
8. **Persuasive feedback:** Stakeholder feedback, progress reports, impact analysis.

In urban development and building projects, the use of smart technology solutions should be a continuous process, starting from ***initial engagement*** with the technological system, requiring easy reachability of the interactive system as well as cognitive ease of using it; proceeding to ***building the engagement*** with interactivity, immersion and personalisation; and creating ***long-term, sustained engagement*** through techniques such as gamification, social interaction and persuasive feedback. Used in carefully designed combinations, these engaging attributes of smart technology can lead to positive emotions, cognitive involvement, increased motivation, and behaviour change in sustainable built environments. Figure 15.1 presents the technology attributes along the phases of engagement.

Figure 15.2 gives an overview of the UrbanEngagingTech framework. The framework is presented as a three-dimensional matrix, where engaging technology attributes are mapped against urban scales and lifecycle phases, with the two sustainability perspectives shaping their implementation. Each cell in the matrix represents potential engagement technologies that can

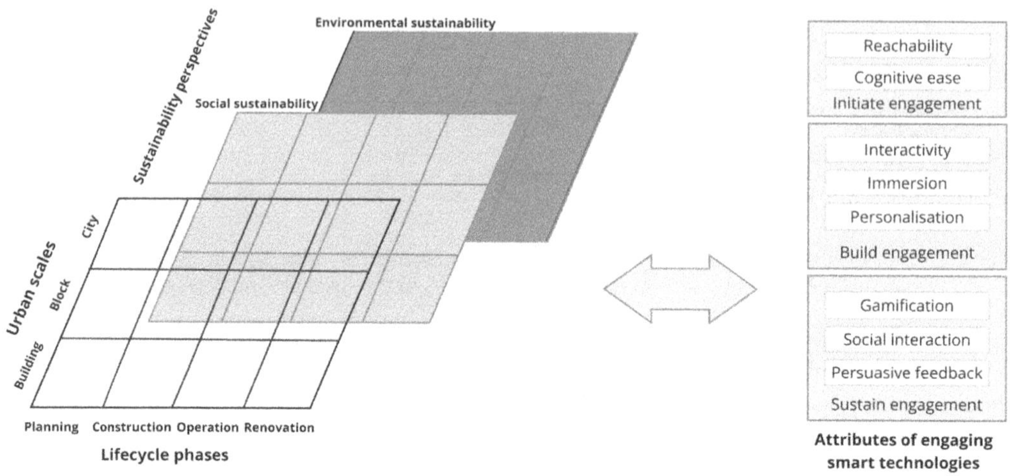

Figure 15.2 The proposed **UrbanEngagingTech** framework. The three dimensions of the framework – the urban scale, lifecycle phase, and sustainability perspective – can be supported by smart technologies that offer means to engage citizens to contribute to sustainable urban development.

address the unique challenges and opportunities at different scales and phases. Below, we outline the key dimensions of the framework.

The framework operates across three urban scales:

- **Building scale** – individual buildings and their residents. Engagement at this scale is typically personal, involving direct user interaction with building technologies and interfaces.
- **Block scale** – clusters of buildings or neighbourhoods, where engagement efforts shift toward collective decision-making and shared infrastructure. This scale supports collaboration among residents, property managers, and other local stakeholders.
- **City scale** – broader urban systems and governance, where citizens engage in city- or district-wide decision-making and large-scale sustainability initiatives.

The framework integrates citizen engagement across four key phases of the urban development lifecycle:

- **Planning and design phase** – initial ideas are formulated, and public input is often sought; current participatory practices predominantly focus on this stage through traditional methods such as public hearings and surveys.
- **Construction phase** – citizen involvement is typically minimal, despite opportunities to engage residents in monitoring and persuasive and experiential feedback.
- **Operation phase** – the use-phase of built urban environments, where citizen participation is crucial for encouraging sustainable behaviours; current approaches rarely facilitate continuous feedback and adaptation.
- **Renewal or renovation phase** – maintaining and repairing buildings covers updates to keep the built environment efficient. This phase is often overlooked in participatory efforts, despite the opportunity to enhance sustainability based on lived experiences.

The framework incorporates two key sustainability perspectives:

- **Social sustainability** – focuses on inclusivity, equity, and communality. Engaging technologies can empower marginalised groups, promote community cohesiveness, and ensure that urban settings reflect the different needs of their inhabitants.
- **Environmental sustainability** – prioritises energy efficiency, resource conservation, and ecological balance. Engaging technologies enable citizens to monitor and help reduce environmental impact and contribute to ecological urban development.

The novelty of the proposed framework lies in its *holistic approach* across different levels and stages of development, and *continuous and adaptive citizen engagement.* Engaging technologies can be tailored to the unique characteristics and needs of each urban design context to ensure their relevance and effectiveness.

Mapping the Case Projects and Engaging Technologies to the UrbanEngagingTech Framework

In this section, we illustrate the dimensions and operationalisation of the UrbanEngagingTech framework with examples from our own case projects (Section "Three Case Project Descriptions and Framework Mapping") and related research (Section "Mapping of Engaging Participatory Technologies to the UrbanEngagingTech Framework").

Three Case Project Descriptions and Framework Mapping

This section describes three case projects in which the development of the built environment is addressed through different urban scales, development lifecycle phases, and the perspectives of environmental and social sustainability. The projects are (1) ***INPERSO,*** focusing on the building scale; (2) ***Nordic Superblocks as Decarbonization Catalysts (NSDC),*** addressing the block scale, and (3) the ***WeGenerate*** project that addresses the broader city and district scale. Table 15.1 gives an overview of the mapping, which is elaborated on in the following subsections, mapping each case project to the different aspects of the framework.

Building Scale – Industrialised and Personalised Renovation for Sustainable Societies (INPERSO) Case Project

INPERSO (https://www.inperso-project.eu/) is an EU-funded project (2023–2026) on inclusive, affordable and sustainable retrofit solutions for existing residential and heritage buildings, demonstrated through pilot studies in the Netherlands, Spain, and Greece. Addressing the entire

Table 15.1 Mapping of the three case projects to UrbanEngagingTech framework

	Urban scale			*Lifecycle phase*				*Sustainability*	
	Building	*Block*	*City*	*Planning*	*Construction*	*Operation*	*Renewal*	*Environmental*	*Social*
INPERSO	x			x	(x)	(x)	x	x	(x)
NSDC		x		x	(x)	x	(x)	x	x
WeGenerate			x	x			x	(x)	x

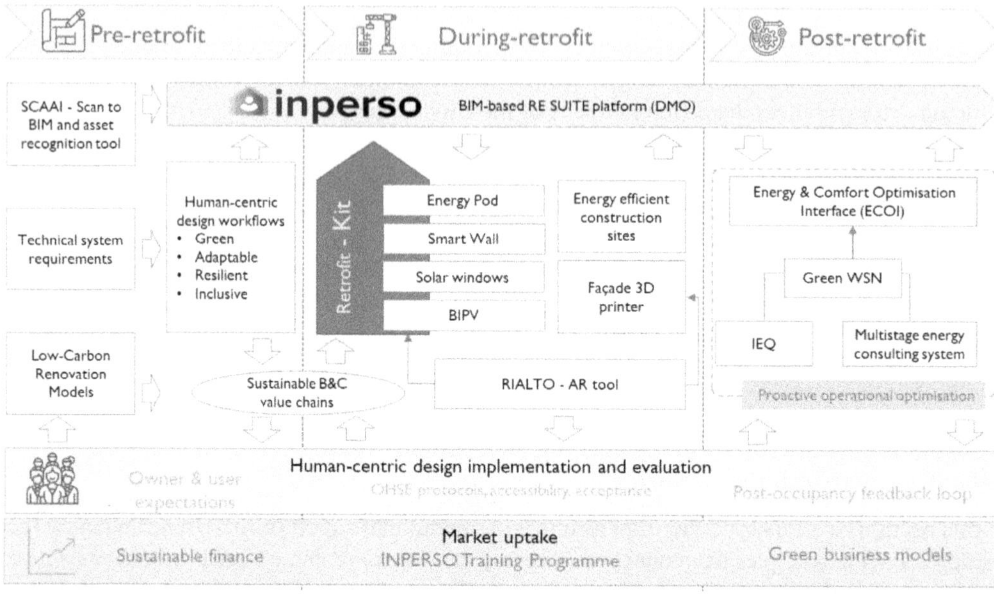

Figure 15.3 Conceptual schematic of the INPERSO project, with a focus on developing digital tools and technologies to accelerate and upscale the transformation of existing buildings to new functions and high sustainability standards.

building lifecycle – from design and renovation to end-of-life – the project integrates smart technologies with human-centred approaches to overcome challenges in digitalisation, fragmented decision-making, quality, efficiency, and rapid retrofit implementation, thereby accelerating and upscaling sustainable renovations (Figure 15.3).

The digital tools of INPERSO include *AI and AR solutions t*o manage building resources and the labour-intensive data collection integrated in a platform called *RE Suite*. This platform streamlines building lifecycle management by reducing data collection inefficiencies. Low-carbon renovation models promote human-centred retrofit design to minimise performance gaps, enhance user satisfaction, and sustainable behaviours. *Retrofit Kit* reduces construction waste, boosts productivity, and enhances workforce skills. Smart innovations include a vertical 3D façade printer to improve retrofit efficiency and reduce carbon footprints. An *energy and comfort optimisation interface* personalises renovation plans, especially for warm climates with natural ventilation, ensuring long-term energy efficiency and comfort.

INPERSO has used workshops and surveys to engage residents and other citizens to understand their perceptions about smart building interactions. Altogether, 18 people with diverse demographic backgrounds participated in three workshops, and 11 homeless people responded to the survey. The findings have been fed back to the participating companies for their consideration in future technology development.

From an *environmental sustainability* perspective, INPERSO aims to cut construction waste, improve energy efficiency, reduce greenhouse gas emissions, and advance low-carbon, resource-efficient renovation via advanced digital and smart construction technologies. *Social sustainability* is pursued through inclusive, human-centred design, increasing digital participation and community resilience.

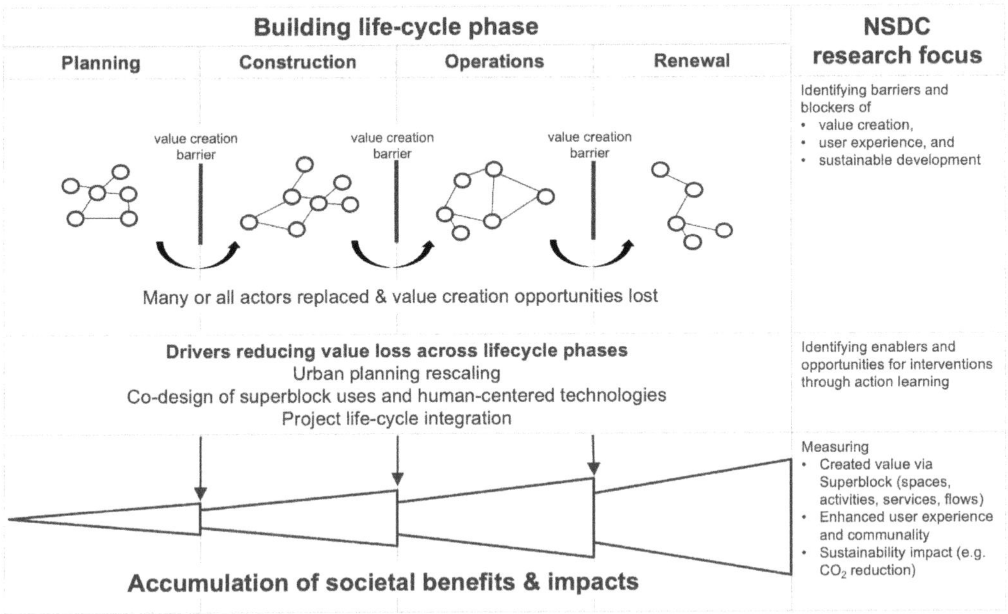

Figure 15.4 The NSDC project addresses all urban development lifecycle phases of the Nordic Superblock by digital tools and close stakeholder collaborations.

Aligned with design attributes for engaging smart technologies, INPERSO enhances *reachability* through accessible tools like *Inhabitant App* for visualising renovation scenarios. *Cognitive ease* is ensured by AI and AR integration, simplifying complex processes. *Immersion* is achieved with AR visualisations for interactive planning. *Persuasive feedback* comes from real-time monitoring in systems like the Retrofit Kit and 3D façade printing. *Personalisation* is central in the energy and comfort interface, tailoring renovations to user needs, while *social interaction* is encouraged via collaborative digital platforms involving diverse stakeholders.

Block Scale: Nordic Superblocks and Decarbonisation Catalysts (NSDC) Case Project

The Nordic Superblocks as Decarbonization Catalysts (NSDC, 2023–2025) project (see https://www.walcc.org/nsdc/), funded by Business Finland, industrial partners, and Tampere University, studies and pilots the Nordic Superblock concept (Sjöblom et al., 2021) and innovative planning, housing, and living models for affordable, sustainable living through shared resources among multiple residential buildings and blocks. This approach emphasises new collaboration forms between stakeholders, including citizens. (Makkonen et al., 2023a).

NSDC adopts an interdisciplinary, systemic perspective on decarbonising buildings across the full lifecycle — from planning and design to construction, operation, and renewal — while preventing value loss caused by stakeholder shifts and knowledge gaps between phases. It enables continuous co-design by diverse stakeholders, including future residents (see Figure 15.4).

Partner companies have developed block-level energy management technologies for facility managers, while resident-level smart energy solutions remain conceptual. Existing systems already provide data on energy consumption and carbon footprints.

The project developed a design framework for advancing neighbourhood communality with smart technologies in Nordic Superblocks (Makkonen et al., in press). NSDC has used interviews (11 residents and 10 experts such as technology developers and city authorities) and co-design workshops (25 participants, with the focus on young people) to understand the design requirements of smart technologies for superblock communality. The findings of these studies have informed the design framework for intelligent technologies in superblocks. The framework guides the interaction design towards facilitating neighbourhood communality. This framework integrates shared physical neighbourhood, social interaction, and social ties, as *social sustainability*, by building a broad collaborative co-creation model that engages stakeholders. *Environmental sustainability* is advanced through the integration of a local energy system within the superblock, optimising energy production as part of the city's infrastructure towards carbon neutrality.

The central design attributes of engaging smart technology to support citizen engagement focus on *reachability* for easy initial engagement, *interactivity* for immediate understanding of the impact of citizens' own activities, *social interaction* for building sense of community, and *persuasive feedback* for interaction between the multiple stakeholders both during the design and operation of the superblocks.

City and District Scale – WeGenerate Case Project

WeGenerate (2023–2027), an EU-funded project, promotes participatory and co-creative urban regeneration, shifting from building *for* people to building *with* people. It empowers city communities to collaboratively develop sustainable, resilient, inclusive, accessible, and aesthetically enriched neighbourhoods.

In Tampere, the project leverages participatory digital twin technology – called the Citiverse – to engage diverse groups, including children, parents, and seniors, in planning the new railway station area. This interactive digital twin lets citizens virtually explore proposed designs, provide feedback, and help ensure outcomes reflect community needs.

The Tampere digital twin supports planning for the station area's redevelopment and raises public awareness. Accessible via an online desktop interface, it allows users to navigate the city model, while participants follow guided narratives and respond to pedestrian experience questions. Beyond navigation, participants can engage creatively with the digital model, envisioning potential future developments for the area. Figure 15.5 presents a screenshot of the Tampere digital twin, including data from IoT sensors that provide real-time traffic conditions for the area.

Participation focuses on physical, digital, and social dimensions of urban renewal, emphasising accessibility, safety, walkability, reachability, service proximity, and environmental quality – addressing social and environmental sustainability and people-centred urban regeneration and diverse community needs. WeGenerate has conducted three co-design workshops with citizens trying out the Digital Twin. There were in total 41 participants, in a broad age range of 10–65 years. The feedback gathered from these sessions is being used to design further versions of the Digital Twin of Tampere area.

The engaging smart technology attributes that are especially prominent in the current version of the City of Tampere Digital Twin are *cognitive ease*, *interactivity* and *immersion*. In the future, developing means for other engaging attributes, especially *reachability*, *social interaction* and *persuasive feedback* within the digital twin would strengthen the impact and engagement of citizens in urban planning and reconstruction.

Figure 15.5 A screenshot of the Tampere digital twin online tool, showcasing a visual representation of the Hervanta area with data collected from traffic-related IoT sensors. (Prototype by Sitowise Inc.)

Mapping of Engaging Participatory Technologies to the UrbanEngagingTech Framework

This section synthesises engaging smart technology examples from our three case studies and additional literature, mapping them to the UrbanEngagingTech framework introduced in Section "The Conceptual Framework for Technology-Mediated Citizen Participation: UrbanEngagingTech". To complement our cases, we conducted a non-systematic literature search in this topic. The goal is to illustrate the variety of engaging smart technologies for citizen participation across the lifecycle phases and scales of the built environment. We present and map 17 examples – 10 supporting social sustainability and 7 supporting environmental sustainability. While we acknowledge overlaps between the two sustainability perspectives, we distinguish them to highlight their specific contributions. Each example and its central engaging attributes are briefly described in Table 15.2 for social sustainability and in Table 15.3 for environmental sustainability. (Some cells are empty because no suitable examples were found.)

Discussion and Conclusion

In this chapter we proposed a new framework, UrbanEngagingTech, using smart technology to engage citizens in the participatory design of urban built environments. Through its potential to move beyond tokenism – superficial inclusion without real influence – of citizen participation, it addresses a critical gap in conventional urban development, which often limits engagement to the planning phase due to its basic methods of surveys or public hearings. Instead, the UrbanEngagingTech framework could enhance systematic and engaging ways for citizen engagement across all different stages of urban development.

Table 15.2 Mapping of the engaging smart technology examples addressing **social sustainability** to the UrbanEngagingTech framework

Phase	*Urban scale*		
	Building	*Block*	*City*
Planning / Design	**IrisVR's Prospect** is a virtual reality platform that enables immersive 3D walkthroughs for architectural and urban design, promoting social sustainability through inclusive, collaborative decision-making. It allows architects, planners, and community members to visualise, experience, and refine building designs together before construction begins (IrisVR, n.d.). **Engaging smart technology attributes:** *cognitive ease*, by simplifying complex designs through an intuitive interface; *immersion,* by offering realistic, full-scale 3D experiences; and *interactivity,* enabling stakeholders to engage in real-time design processes.	**Urban Data Platform (UDP) and Array of Things (AoT)** use IoT devices and sensors to collect real-time urban and environmental data, enabling citizen engagement in planning and sustainability efforts. The UDP integrates data across city systems to support interoperable services, and empowers citizens with access to information and feedback tools. The AoT sensor network captures environmental and personal data, allowing users to monitor urban conditions (Argonne National Laboratory, n.d.). **Engaging smart technology attributes:** r*eachability*, by offering accessible data and feedback tools; *interactivity,* through real-time engagement with urban conditions; and *persuasive feedback,* enabling transparent citizen influence on urban decisions.	**Tampere Digital Twin** uses an interactive digital twin model to involve diverse citizen groups in the planning and development of the new railway station area – see Section "City and District Scale – WeGenerate Case Project". **Engaging smart technology attributes:** see Section "City and District Scale – WeGenerate Case Project".
Construction	**Wikihouse and 3D printing technology** support social sustainability at the building scale by democratising production and promoting inclusivity in urban development. By transforming architecture into an open-source, citizen-driven process, they enable anyone with internet access and basic fabrication tools to design and construct their		**CityScopeAR** is a mixed-reality platform for collaborative urban design, enabling stakeholders to visualise and interact with projects in real time, supporting social and environmental sustainability through transparent decision-making and adaptive design. By combining AR, real-

Table 15.2 (Continued)

Phase	*Urban scale*		
	Building	*Block*	*City*
	own homes (UIC Barcelona, 2013). **Engaging smart technology attributes:** *reachability,* through broad access to design and construction resources; *cognitive ease,* by offering intuitive open-source tools; and *interactivity,* which empowers users to engage hands-on in shaping their living environments.		time simulation, and interactive modelling, it allows users to explore scenarios and provide input before construction begins (Noyman et al., 2018). **Engaging smart technology attributes:** *cognitive ease*, achieved by translating complex data into intuitive visuals; *interactivity,* through real-time exploration and modification of urban models; and *persuasive feedback*, which delivers instant insights to support collaborative decisions.
Operation		**Commu** is a map-based online help-exchange application that enables residents of specific communities to ask for and offer assistance, fostering support networks and reducing loneliness through practical task exchange and social interaction (Makkonen et al., 2023b). By offering personalised recommendations based on proximity and type of help, it strengthens neighbourhood ties and promotes communality. **Engaging smart technology attributes:** *social interaction*, as the platform facilitates easy and recurring connections; *reachability*, by making help accessible within local communities; and *personalisation,* through tailored recommendations to enhance user engagement.	**The Multimodal Ambient Context-enriched Intelligence Platform (MACeIP)** is a smart city system that enhances urban management and citizen engagement by integrating IoT sensors, edge computing, multimodal AI, and cloud services. It offers a comprehensive, interactive approach to real-time urban decision-making, with a prototype implemented in Fredericton, New Brunswick, Canada (Nguyen et al., 2024). **Engaging smart technology attributes:** *immersion,* enabling users to deeply engage with the system; *interactivity,* through real-time feedback and decision-shaping; and *persuasive feedback,* allowing adaptive responses based on citizen input.

(*Continued*)

Table 15.2 (Continued)

Phase	*Urban scale*		
	Building	*Block*	*City*
Renovation / Renewal	**The Inhabitant App** is a digital platform that enables building occupants to provide feedback during the renovation process, promoting social sustainability by ensuring resident needs and preferences are reflected in planning decisions (INPERSO, Section "Building Scale – Industrialised and Personalised Renovation for Sustainable Societies (INPERSO) Case Project"). **Engaging smart technology attributes:** *reachability*, by offering easy access to real-time information and direct engagement with stakeholders; *interactivity*, enabling active citizen participation in shaping their living environment; and *persuasive feedback*, which ensures that user input meaningfully influences renovation outcomes.	**UrbanistAI** is a generative AI platform for participatory co-design and visualisation that empowers communities to collaboratively reimagine and improve social places at the block scale. By enabling real-time simulations and visual tools, it supports inclusive and transparent planning processes that promote socially sustainable, adaptable urban spaces (UrbanistAI, n.d.). **Engaging smart technology attributes:** *reachability*, through accessible tools that invite broad community participation; *cognitive ease*, by simplifying complex planning through intuitive, data-driven simulations; and *interactivity*, allowing real-time engagement with visualisations and collaborative design processes.	**PlaceSpeak** is a location-based civic engagement platform that connects citizens with local issues to support inclusive, transparent, and representative urban renewal efforts (PlaceSpeak, n.d.). It uses secure authentication to verify users by location, ensuring relevance and equity in participation. **Engaging smart technology attributes:** *persuasive feedback*, through geospatial polling and real-time input that amplifies local voices; *cognitive ease*, by offering an intuitive, location-based interface; and *interactivity*, enabling citizens to actively shape urban renewal through ongoing digital engagement.

Some of the engaging technology attributes for participatory engagement are more prominent than others, as presented in Section "Mapping the Case Projects and Engaging Technologies to the UrbanEngagingTech Framework". This review of existing emerging technology systems highlighted that *cognitive ease* and *interactivity* are essential to any engaging participatory systems. Some systems also offer *immersion*, e.g., through VR, some form of motivating *gamification* elements to build and sustain engagement. However, what seems to be lacking in many of the systems is easy *reachability* to the participatory applications, i.e., access to them "anytime, anywhere". Also typically missing are *personalisation* (i.e., allowing the system to adapt to users' preferences and behaviour), *social interaction* among community members, and *persuasive feedback* from the urban stakeholders directly through the system. Inclusion of these missing aspects in engaging smart technologies could be a key to long-term citizen engagement in urban development and associated behaviour change.

Table 15.3 Mapping of the engaging smart technology examples addressing **environmental sustainability** to the UrbanEngagingTech framework

Phase	*Urban scale*		
	Building	*Block*	*City*
Planning / Design			**Helsinki 3D+'s Energy and Climate Atlas** is a 3D-based tool that promotes environmental sustainability by providing transparent data on building energy use and renewable energy potential, such as solar and geothermal resources. It empowers citizens, planners, and businesses to make informed decisions and contribute to climate goals (City of Helsinki, 2024). **Engaging smart technology attributes:** *cognitive ease,* through intuitive 3D visualisations that simplify complex environmental data; and *interactivity*, allowing real-time exploration of energy consumption and renewable potential.
Construction		**Virtual environment of Asemanranta,** Hämeenlinna is a 3D platform for visualising an area under construction along with citizen-relevant data. It integrates building models, zoning information, district heating networks, and open data such as maps, railway infrastructure, energy use, soil conditions, and population metrics (Kukkamäki, 2023). **Engaging smart technology attributes:** *cognitive ease,* by intuitively presenting complex zoning and environmental data; *immersion,* through interactive 3D visualisations that simulate future neighbourhoods; and *interactivity,* allowing real-time exploration of infrastructure.	

(*Continued*)

Table 15.3 (Continued)

Phase	*Urban scale*		
	Building	*Block*	*City*
Operation	**GreenLife – social robot** is an application concept using the social robot Pepper to promote sustainable behaviour among residents in shared spaces. By delivering sustainability information, offering rewards, and providing social feedback, the robot encourages energy saving, waste reduction, and responsible resource use through an engaging, community-oriented approach (Beheshtian et al., 2020). **Engaging smart technology attributes:** *persuasive feedback,* through personalised behavioural suggestions; *gamification,* using rewards to motivate eco-friendly actions; and *immersion,* achieved via the robot's physical presence and social engagement.	**Peer-to-peer sharing of energy storage systems** enables households to locally buy, sell, and exchange stored electricity using net metering and time-of-use pricing. By leveraging cooperative game theory and dynamic pricing, the system ensures fair energy distribution and cost efficiency, while advancing sustainability through greater renewable energy use and reduced carbon emissions (Victor Sam Moses Babu et al., 2023). **Engaging smart technology attributes:** *interactivity,* allowing users to manage storage and monitor usage in real time; *persuasive feedback,* nudging sustainable energy behaviours; and *gamification,* motivating optimisation through incentives and shared goals.	**Waste Recycling Chatbots** use AI-driven guidance to educate and assist citizens in local waste management, promoting environmental sustainability by reducing landfill waste, lowering emissions, and conserving resources. They enhance the efficiency of recycling systems and foster broader citizen engagement through accessible digital platforms (Zota et al., 2024). **Engaging smart technology attributes:** *reachability,* by offering city-wide real-time support; *interactivity,* enabling instant feedback and questions; and *persuasive feedback,* reinforcing behavioural change toward responsible waste disposal.
Renovation / Renewal		**The Green Kalasatama AR application** engages citizens in green space planning by allowing them to customise future landscapes in developing urban areas. Through a mobile AR platform, users can add trees and plants to a virtual environment, while receiving information about their benefits. The tool supports collaboration between residents and planners, enhancing shared decision-making in urban greening (Fiksukaupunki, 2022). **Engaging smart**	**Cloudburst's GIS and interactive platforms** in Copenhagen support climate resilience by engaging citizens in co-designing green infrastructure and flood mitigation strategies. Through accessible digital tools, residents can map flood-prone areas, propose sustainable interventions, and collaborate with planners, fostering environmental sustainability, trust, and shared ownership in urban planning (Berigüete et al., 2024).

Table 15.3 (Continued)

Phase	*Urban scale*		
	Building	*Block*	*City*
		technology attributes: *interactivity*, enabling users to actively design virtual landscapes; *personalisation,* adapting the experience to individual and neighbourhood needs; and *reachability,* through its accessible, mobile-based platform.	**Engaging smart technology attributes:** *interactivity,* enabling real-time citizen input; *persuasive feedback,* reinforcing users' contributions to resilience; and *personalisation,* adapting tools to neighbourhood-specific concerns.

While emerging technologies can offer powerful tools for participatory urban planning, several barriers may hinder their widespread and equitable use. Issues of equity and accessibility remain significant, as advanced platforms may inadvertently reinforce existing inequalities. To support inclusive engagement, hybrid approaches are essential, e.g., combining high-tech tools like AR and generative AI with low-tech methods such as physical materials (Fredericks et al., 2018). This must go hand-in-hand with human-centred design for digital literacy and for diverse user needs to help bridge the digital divide. Equally important is transparent governance, including citizen boards and real-time feedback loops because this is crucial for building trust and ensuring community voices are genuinely reflected in decision-making (Pietilä et al., 2021).

Finally, maintaining long-term participation requires addressing behavioural and social dynamics, not just technological ones. Short-term incentives such as gamification may encourage initial engagement, but lasting change is more likely when technology is combined with social structures and sense of ownership through co-creation workshops and community rewards. At the same time, data privacy and security must be addressed so that citizens have full transparency and control over how their data is collected and used. This trust in participatory processes is a delicate issue, and the applied citizen technologies must be designed to enforce that trust.

Conclusion

The design of built urban environments should adopt a holistic approach that integrates citizens' perspectives with the views of other key stakeholders, such as city authorities, urban planners, and construction companies. This is because all citizens should offer their local, experience-based views and needs about their lived urban environments and sustainability issues at different development stages. However, current practices often lack long-term mechanisms for such holistic citizen participation in sustainable development.

The UrbanEngagingTech framework is proposed as a methodological contribution to technology-mediated urban design for two key audiences: technology developers designing digital tools for urban participation, and urban developers seeking to enhance long-term citizen engagement throughout the development lifecycle. The attributes of engaging technologies intend to contribute to a more inclusive future-built environment by holistically addressing the smart technology-enabled ways to involve citizens. Inclusion of digital tools with the guidance of the framework could lead to increased citizen motivation, emotional experiences, and long-term behaviour change. Furthermore, the framework can guide urban officials and stakeholders in understanding novel ways of smart technology solutions for participatory urban approaches.

The UrbanEngagingTech framework was developed through iterative literature synthesis and analysis of existing case examples; future research should empirically evaluate the framework's practical applicability in diverse urban contexts.

References

AlWaer, H., & Cooper, I. (2020). Changing the focus: Viewing design-led events within collaborative planning. *Sustainability*, *12*(8), 3365. https://doi.org/10.3390/SU12083365

Argonne National Laboratory. (n.d.). Array of Things. www.anl.gov/mcs/array-of-things

Beheshtian, N., Moradi, S., Ahtinen, A., Väänänen, K., Kähkonen, K., & Laine, M. (2020). Greenlife: A persuasive social robot to enhance the sustainable behavior in shared living spaces. *Proceedings of the 11th Nordic Conference on Human-Computer Interaction: Shaping Experiences, Shaping Society*, 1–12. https://doi.org/10.1145/3419249.3420143

Berigüete, F. E., Santos, J. S., & Rodriguez Cantalapiedra, I. (2024). Digital revolution: Emerging technologies for enhancing citizen engagement in urban and environmental management. *Land*, 13(11), 1921. https://doi.org/10.3390/land13111921

Bibri, S. E. (2020). Data-driven environmental solutions for smart sustainable cities: Strategies and pathways for energy efficiency and pollution reduction. *Euro-Mediterranean Journal for Environmental Integration*, 5(3), 66. https://doi.org/10.1007/s41207-020-00211-w

Bibri, S. E., Alexandre, A., Sharifi, A., & Krogstie, J. (2023). Environmentally sustainable smart cities and their converging AI, IoT, and big data technologies and solutions: An integrated approach to an extensive literature review. *Energy Informatics*, *6*(1), 9. https://doi.org/10.1186/s42162-023-00259-2

Castaño-Rosa, R., Pelsmakers, S., Wang, J., & Cao, S. (2024). Building management systems in residential buildings: Their role in energy and indoor climate resilience. *Architectural Research in Finland*, *8*(1), 42–62. https://doi.org/10.37457/arf.146873

Chado, J., Johar, F. B., & Khan, M. G. (2017). Citizen participation in urban governance: Experiences from traditional city of Bida, Nigeria. *Advanced Science Letters, 23*(7), 6194–6197. https://doi.org/10.1166/asl.2017.9234

City of Helsinki. (2024). Helsinki 3D. Retrieved February 5, 2025, from www.hel.fi/en/decision-making/information-on-helsinki/maps-and-geospatial-data/helsinki-3d#urban-data-model

Dempsey, N., Bramley, G., Power, S., & Brown, C. (2011). The social dimension of sustainable development: Defining urban social sustainability. *Sustainable Development, 19*(5), 289–300. https://doi.org/10.1002/sd.417

Fagerholm, N., Raymond, C. M., Olafsson, A. S., Brown, G., Rinne, T., Hasanzadeh, K., Broberg, A., & Kyttä, M. (2021). A methodological framework for analysis of participatory mapping data in research, planning, and management. *International Journal of Geographical Information Science, 35*(9), 1848–1875. https://doi.org/10.1080/13658816.2020.1869747

Fanani, A. Z., Hastuti, K., Syarif, A. M., & Harsanto, P. W. (2021). Challenges in developing virtual reality, augmented reality and mixed-reality applications: Case studies on a 3D-based tangible cultural heritage conservation. *International Journal of Advanced Computer Science and Applications*, *12*(11), 219–227. https://doi.org/10.14569/ijacsa.2021.0121126

Fiksukaupunki. (2022). Green Kalasatama. Retrieved March 3, 2025, from https://fiksukaupunki.fi/en/projects/green-kalasatama/

Foroughi, M., Wang, T., & Pereira Roders, A. (2025). In praise of diversity in participatory heritage planning empowered by artificial intelligence: Windcatchers in Yazd. *Urban Planning, 10*(S2), Article 8724. https://doi.org/10.17645/up.8724

Fredericks, J., Hespanhol, L., Parker, C., Zhou, D., & Tomitsch, M. (2018). Blending pop-up urbanism and participatory technologies: Challenges and opportunities for inclusive city making. *City, Culture and Society*, *12*, 44–53. https://doi.org/10.1016/j.ccs.2017.06.005

Fredericks, J., Tomitsch, M., & Haeusler, M. H. (2020). Redefining community engagement in smart cities: Design patterns for a smart engagement ecosystem. In C. Silva (Ed.), *Citizen-Responsive Urban E-Planning: Recent Developments and Critical Perspectives* (pp. 13–53). IGI Global Scientific Publishing. https://doi.org/10.4018/978-1-7998-4018-3.ch002

Fredricks, J. A., Blumenfeld, P. C., & Paris, A. H. (2004). School engagement: Potential of the concept, state of the evidence. *Review of Educational Research*, *74*(1), 59–109. https://doi.org/10.3102/00346543074001059

Fung, A. (2015). Putting the public back into governance: The challenges of citizen participation and its future. *Public Administration Review, 75*(4), 513–522. https://doi.org/10.1111/puar.12361

Gholami, M., Torreggiani, D., Barbaresi, A., & Tassinari, P. (2024). Smart green planning for urban environments: The city digital Twin of Imola. In F. Belaïd & A. Arora (Eds.), *Smart Cities: Social and Environmental Challenges and Opportunities for Local Authorities* (pp. 133–150). Springer International Publishing. https://doi.org/10.1007/978-3-031-35664-3_10

Hadiyana, T., & Ji-hoon, S. (2024). AI-driven urban planning: Enhancing efficiency and sustainability in smart cities. *ITEJ (Information Technology Engineering Journals*), *9*(1), 23–35. https://doi.org/10.24235/itej.v9i2.124

Hochheiser, H., & Lazar, J. (2007). HCI and societal issues: A framework for engagement. *International Journal of Human–Computer Interaction, 23*(3), 339–374. https://doi.org/10.1080/10447310701702717

IrisVR. (n.d.). *Prospect: Immersive VR Walkthroughs.* Retrieved February 23, 2025, from https://irisvr.com/prospect

Jamei, E., Mortimer, M., Seyedmahmoudian, M., Horan, B., & Stojcevski, A. (2017). Investigating the role of virtual reality in planning for sustainable smart cities. *Sustainability*, *9*(11), 2006. https://doi.org/10.3390/su9112006

Kaewkitipong, L., Chen, C., Han, J., & Ractham, P. (2022). Human–computer interaction (HCI) and trust factors for the continuance intention of mobile payment services. *Sustainability*, *14*(21), 14546. https://doi.org/10.3390/su142114546

Khan, Z., Dambruch, J., Peters-Anders, J., Sackl, A., Strasser, A., Fröhlich, P., Templer, S., & Soomro, K. (2017). Developing knowledge-based citizen participation platform to support smart city decision making: The smarticipate case study. *Information, 8*(2), 47. https://doi.org/10.3390/info8020047

Kukkamäki, J. (2023). *Virtuaaliympäristöt kaupunkiympäristön suunnittelussa ja yhteiskehittämisessä.* [Master's thesis, Tampere University]. Trepo. https://urn.fi/URN:NBN:fi:tuni-202305155755

Makkonen, J. J., Hartikainen, M., Nygren, R., & Väänänen, K. (2023b). User perceptions of help posts in commu help exchange application. *Proceedings of the 26th International Academic Mindtrek Conference*, 329–333. https://doi.org/10.1145/3616961.3617809

Makkonen, J., Kaipainen, K., Laine, M., & Väänänen, K. (in press). Advancing neighbourhood communality with intelligent technologies in Nordic superblocks. In D. Cushing, V. Garcia Hansen, G. Amayo Caldwell, & M. Foth (Eds.), *Human-Building Interaction: The Nexus of Architecture, Building Science & Interaction* Design. Springer.

Makkonen, J., Latikka, R., Kaukonen, L., Laine, M., & Väänänen, K. (2023a). Advancing residents' use of shared spaces in Nordic superblocks with intelligent technologies. *AI & Society, 38*(3), 1167–1184. https://doi.org/10.1007/s00146-022-01604-x

McCarthy, J., & Wright, P. (2005). Putting 'felt-life' at the centre of human–computer interaction (HCI). *Cognition, Technology & Work*, *7*(4), 262–271. https://doi.org/10.1007/s10111-005-0011-y

Næss, P. (2001). Urban planning and sustainable development. *European Planning Studies, 9*(4), 503–524. https://doi.org/10.1080/713666490

Nguyen, T. T. H., Nguyen, P. T. L., Wachowicz, M., & Cao, H. (2024). MACeIP: A multimodal ambient context-enriched intelligence platform in smart cities. *2024 IEEE International Conference on Consumer Electronics-Asia (ICCE-Asia)*, 1–4. https://doi.org/10.1109/ICCE-Asia63397.2024.10774046

Noyman, A., Sakai, Y., & Larson, K. (2018). CityScopeAR: Urban design and crowdsourced engagement platform. *Proceedings of the CHI Conference on Human Factors in Computing Systems* (pp. 121–137). ACM. https://doi.org/10.13140/RG.2.2.11178.67528

O'Brien, H. L., & Toms, E. G. (2008). What is user engagement? A conceptual framework for defining user engagement with technology. *Journal of the American Society for Information Science and Technology, 59*(6), 938–955. https://doi.org/10.1002/asi.20801

Oinas-Kukkonen, H., & Harjumaa, M. (2009). Persuasive systems design: Key issues, process model, and system features. *Communications of the Association for Information Systems, 24*, 485–500. https://doi.org/10.17705/1CAIS.02428

Pietilä, I., Meriläinen, N., Varsaluoma, J., & Väänänen, K. (2021). Understanding youths' needs for digital societal participation: Towards an inclusive virtual council. *Behaviour & Information Technology, 40*(5), 483–496. https://doi.org/10.1080/0144929X.2021.1912182

PlaceSpeak. (n.d.). PlaceSpeak. Retrieved February 25, 2025, from www.placespeak.com/en/

Prestopnik, N., & Zhang, P. (2015). Human–computer interaction (HCI): Interactivity, immersion, and invisibility as new extensions. In *Wiley Encyclopedia of Management* (pp. 1–6). John Wiley & Sons, Ltd. https://doi.org/10.1002/9781118785317.weom070023

Rapp, A., Hopfgartner, F., Hamari, J., Linehan, C., & Cena, F. (2019). Strengthening gamification studies: Current trends and future opportunities of gamification research. *International Journal of Human-Computer Studies, 127*, 1–6. https://doi.org/10.1016/j.ijhcs.2018.11.007

Sharma, N. K., Hargreaves, T., & Pallett, H. (2023). Social justice implications of smart urban technologies: An intersectional approach. *Buildings and Cities, 4*(1), 315–333. https://doi.org/10.5334/bc.290

Sjöblom, J., Kuoppa, J., Laine, M., & Alatalo, E. (2021). Crafting a planning issue with citizens in the context of planning competition: A case of "Nordic superblock". *Journal of Urban Design, 26*(1), 117–131. https://doi.org/10.1080/13574809.2020.1832886

UIC Barcelona. (2013, May 24). The WikiHouse: Housing for the 100 percent? Master of international cooperation sustainable emergency architecture. Retrieved April 3, 2025, from https://masteremergencyarchitecture.uic.es/2013/05/24/the-wikihouse-an-answer-to-the-worlds-housing-problems/

UrbanistAI. (n.d.). UrbanistAI. Retrieved January 10, 2025, from https://site.urbanistai.com

Victor Sam Moses Babu, K., Satya Surya Vinay, K., & Chakraborty, P. (2023). Peer-to-peer sharing of energy storage systems under net metering and time-of-use pricing. *IEEE Access, 11*, 3118–3128. https://doi.org/10.1109/access.2023.3234625

Wilson, A., Tewdwr-Jones, M., & Comber, R. (2019). Urban planning, public participation and digital technology: App development as a method of generating citizen involvement in local planning processes. *Environment and Planning. B, Urban Analytics and City Science, 46*(2), 286–302. https://doi.org/10.1177/2399808317712515

Yigitcanlar, T., Desouza, K. C., Butler, L., & Roozkhosh, F. (2020). Contributions and risks of artificial intelligence (AI) in building smarter cities: Insights from a systematic review of the literature. *Energies (Basel), 13*(6), 1473. https://doi.org/10.3390/en13061473

Yigitcanlar, T., Kamruzzaman, M., Foth, M., Sabatini-Marques, J., da Costa, E., & Ioppolo, G. (2019). Can cities become smart without being sustainable? A systematic review of the literature. *Sustainable Cities and* Soci*ety, 45,* 348–365. https://doi.org/10.1016/j.scs.2018.11.033

Zota, R. D., Cîmpeanu, I. A., Dragomir, D. A., & Lungu, M. A. (2024). Practical approach for smart and circular cities: Chatbots used in waste recycling. *Applied Sciences, 14*(7), 3060. https://doi.org/10.3390/app14073060

Conclusion

16 Conclusions

Synthesizing Participatory Design, Engagement, and Innovation for the Future

Ravi Mahamuni, Jonna Häkkilä, and Mari Suoheimo

Introduction

This final concluding chapter brings together the main insights from all the book chapters. It aims to provide a clear and accessible synthesis of the themes addressed in the book, providing summarizing insights that are useful for academics, practitioners, and policymakers. The goal is to highlight cross-cutting lessons, reflect on challenges, and suggest next steps for research and practice. The different themes can come together to improve civic engagement, more demographic organizational change and decision making, responsible innovation, and more sustainable futures through increased engagement of the stakeholders and participatory design.

The focus of the book has been on participatory design, engagement, and forward-looking innovation across diverse sectors. We can see that participatory design, which involves engaging people in designing systems relevant to themselves, is a global phenomenon, and the case studies demonstrate how these approaches can be implemented regardless of the location. Technologies can be part of the innovation process alongside the people and non-human agents. In this way, technologies can be both the enablers and the end result of the process, in the form of a product or a service.

Looking at the Landscapes of Engagement and Participatory Design

When thinking of the general landscape of engagement and participatory design, the vastness of the field is probably the first notion to remark. As participatory design already builds on a long-standing tradition (Gregory, 2003) and sustains a biannual academic conference to foster a research community (D'Andrea et al., 2024), the body of research falling under the methodological approach is large. To form an overview of the landscape in any certain research field, a literature review that systematically goes through the prior art is a good approach. This has been conducted in Chapter 2, which brings forward the findings of a scoping review in the cross-section of participatory action research (PAR), co-design, engagement, and technology, by screening a total of 231 studies and the final dataset of 32 research articles. PAR serves as a guiding framework for co-design, relying on cycles of planning, action, and review.

In prior art, digital platforms and multimedia tools emerged as central to participatory processes and engagement, as reported in Chapter 2. They help participants share ideas and provide feedback, ensuring that solutions are relevant to daily life. The literature review also revealed common challenges in implementing PAR. These include difficulties in achieving genuine collaboration, the time-consuming nature of the process, recruitment challenges, and bridging PAR outcomes to policy-making. In the larger landscape of participatory design, these challenges are something to be addressed in the future, opening an opportunity where

DOI: 10.4324/9781003667872-21

we can seek help from technology-assisted methods. An example of a new type of approach in engaging citizens and stakeholders, Chapter 3 introduced the Living Law Lab as an example of how participatory approaches extend beyond product or service design. By linking law, AI, and design, the Lab offers a model for questioning systemic exclusions and creating infrastructures of accountability and care. Addressing the emerging use of AI in society, Chapter 3 brings up the importance of participatory design in a role, which can minimize harms, question the direction of algorithm development, and push for accountability when developing legislation.

Engagement in Shaping Democracy

Chapters 4–7 explored how participatory design can reshape democratic practices. They push for hearing people's voices and empowering the citizens to influence important societal issues such as budget allocation, safety, and environmental issues. Democracy can be strengthened by these participatory actions, and it has been pointed out that listening can help in the democratic objectives of enhancing legitimacy, helping to deal with disagreements, improving understanding, and increasing empowerment (Dobson, 2012). Chapters 4–7 offer insight into how listening to the people can be realized in practice and provide guidance and lessons learned for future practitioners who can apply the same methods. Together, these chapters show how participatory design can support democratic innovation.

Chapter 4 examined Boston's participatory budgeting initiative, inspired by Latin American experiences, and showed how inclusive processes redistribute agency and build trust. The participatory budgeting process explored using both material and digital design elements for engaging citizens and expanding participation, resulting in a practical example for sharing civic power. Chapter 5 introduced the SafetySpinner tool for the tourism industry, demonstrating how participatory dialogue can highlight safety concerns and strengthen service quality. Chapter 6 showed how generative AI can stimulate discussion about local futures in Arctic communities. Images generated with community members supported intergenerational dialogue and futures literacy while also raising ethical questions about representation. Chapter 7 proposed a broader vision of civic engagement, arguing that citizen-centered design thinking and gamified, technology-mediated platforms can lower barriers for participation and make engagement more inclusive. Utilizing games and social media for citizen participation makes use of channels and tools that are already familiar to people. Harnessing everyday behaviors, such as sharing on social media and voicing complaints, can bring those previously disengaged to become informants for policy-making.

Holistic Sustainability and Way of Living

Sustainable ways of living and design solutions that support this have become ever more important with the world facing a risk of ecological crisis. A sustainable future is calling for actions that promote behavior change, demand smaller consumption of natural resources, consider also more-than human actors in design. At the same time, social sustainability is increasingly important in order to maintain functioning societies. Chapters 8–11 highlighted the role of participatory design in advancing sustainability and responsible innovation.

Chapter 8 examined digital health innovations in China, showing that responsible participatory design requires engagement across the entire lifecycle of technology. Incorporating anticipation, reflexivity, and responsiveness leads to stronger adoption and long-term impact. Chapters 9 and 10 addressed the development of recycling and reusing services. Chapter 9 investigated a deposit-return system for reusable shopping bags through experience prototyping, demonstrating how service infrastructures shape engagement. Prototyping and user feedback revealed both potential and barriers, underscoring the importance of aligning interventions with everyday habits. Chapter 10 extended this theme with an in-the-wild study involving two airport restaurants and QR-coded reusable bowls with a supporting infrastructure for reuse. The findings highlighted that behavioral, institutional, and contextual factors must be addressed alongside technological solutions when designing reuse systems. Chapter 11 expanded the scope of sustainability to consider a more planetary view and non-human actors, exploring participatory approaches in zoo design with humans, animals, and AI. The project showed how interspecies engagement can inspire ethical and sustainable practices.

Altogether, sustainability is a recurring theme in the book. Collectively, Chapters 8–11 argue for participatory approaches that are reflective, inclusive, and attentive to both human and more-than-human stakeholders. The studies reveal that sustainable design interventions tend to be more successful if they fit into daily life and can integrate with existing routines and behavioral models. Also, the introduction of systems that support more sustainable choices, such as reusing products, requires considering the whole service ecosystem related to the product.

Future Smart Cities and Transportation

Adopting technologies has shaped the everyday life of individuals and communities throughout human history. Development of urban environments and commuting infrastructure are key areas for the decision-makers to consider, and technological solutions play a major role in how these futures will look. We can expect new challenges and opportunities to arise as smart city infrastructures and autonomous physical agents such as self-driving cars and robots emerge. Chapters 12–15 turned to future urban environments and mobility and engaging people in envisioning their future.

Chapter 12 described a participatory foresight study in Rio de Janeiro, which engaged citizens in envisioning mobility futures and generated hundreds of actionable proposals. The research highlighted both the value of personal relevance in participation and the importance of including institutional actors to ensure impact. Chapters 13 and 14 utilize technologies that are expected to become commonly used in the future, robots and virtual production stages, for exploring what urban futures could look like. Chapter 13 examined human–robot interaction in different settings in the wild, showing that acceptance of robot dogs varies by context, from airports to cafés, and that roles such as security or cleaning are perceived differently depending on the setting.

Chapter 14 introduced immersive theater and worldbuilding on virtual production stages as novel methods for prototyping near-future services. By situating participants in speculative scenarios, these approaches supported empathic engagement and interdisciplinary collaboration. Finally, Chapter 15 presented the UrbanEngagingTech framework, which embeds citizen participation across the lifecycle of urban environments and can provide a systematic approach for leveraging technology for more democratic, sustainable, and resilient urban futures. Technologies such as AI, IoT, and VR/AR can move citizens from passive informants to active co-creators, provided that equity, transparency, and trust are prioritized.

Synthesis: Cross-Cutting Lessons and Future Directions

The book reveals several important lessons about participatory design, engagement, and innovation. We are highlighting the following points:

Active engagement is the key. Participation must be meaningful, sustained, and adapted to local contexts. Participatory frameworks must be adaptable and responsive to local needs and constraints. Combining top-down support, bottom-up initiatives, and peer-to-peer facilitation are most effective. It is also important to keep in mind the possibilities for expanding the participation, as including versatile stakeholders, such as non-human actors, can open new directions for design.

Technology is a tool, not a solution. Digital tools and AI can support engagement but only when aligned with human values, cultural contexts, and systemic strategies. Technology should facilitate co-creation rather than replace it. Co-creation, reflexivity, and cultural sensitivity are essential in avoiding the reinforcement of existing exclusions and biases. When technology is used as a tool for engagement, stakeholders' experience with it and the current behavior models and usage practices should be taken into account.

Democracy needs new forms of engagement. Participatory design can strengthen civic trust and accountability. The processes should be collaborative and open to diverse perspectives. Engaging citizens in meaningful ways is beyond procedural requirements, and technology can be harnessed to provide new ways for discussion and co-design. Methods such as generative AI, immersive theater, and worldbuilding offer creative ways to connect imagination with action and bridge the gap between vision and implementation.

Responsibility goes beyond innovation. Responsible design requires attention to the entire lifecycle of technologies and infrastructures. Moreover, a change for a more sustainable product often requires considering the whole service ecosystem around the product. Sustainability depends not only on technical fixes but also on behavioral, institutional, and cultural change. In order to efficiently adopt a more sustainable lifestyle, sustainable systems and solutions should fit into the existing everyday life routines.

Looking ahead, there are several areas where further research is needed. More studies are required to test participatory frameworks in diverse contexts, particularly where barriers to engagement remain high. Stronger integration of institutional actors together with citizen participants can help bridge the gap between visionary ideas and implementation. With continuously advancing technology, including the rapid rise of AI, ethical considerations around inclusivity must stay central. More-than-human perspectives should be further developed to shift from human-centered to planet-centered design in order to create holistically sustainable solutions. Creative and embodied methods, such as immersive prototyping, for strengthening engagement and participation also deserve further exploration.

In conclusion, this book highlights that participatory design and engagement are not static techniques but evolving practices that demand reflection, adaptation, and commitment to shared values. By learning across sectors such as law, health, sustainability, and mobility, we can create more just, inclusive, and innovative systems that better serve societies as a whole.

We hope you have enjoyed reading this book.

The editorial team,
Mari Suoheimo, Jonna Häkkilä and Ravi Mahamuni

Acknowledgments

This work has been partially supported by the project Reconceptualizing Boundaries Together Towards Resilient and Just Arctic Future(s) (REBOUND), grant numbers 358482 and 358498, funded by the Strategic Research Council (SRC) established within the Research Council of Finland, and ProtoXR project funded by ERDF.

References

D'Andrea, V., Abreu de Paula, R., Geppert, A. A., Brereton, M., Gaudio, C. D., Jensen, M. Y., Winschiers-Theophilus, H. & Zaman, T. (Eds.). (2024). *Proceedings of the Participatory Design Conference 2024: Full Papers-Volume 1*. ACM.

Dobson, A. (2012). Listening: The new democratic deficit. *Political Studies*, *60*(4), 843–859.

Gregory, J. (2003). Scandinavian approaches to participatory design. *International Journal of Engineering Education*, *19*(1), 62–74.

Index

For Product Safety Concerns and Information please contact our EU
representative GPSR@taylorandfrancis.com
Taylor & Francis Verlag GmbH, Kaufingerstraße 24, 80331 München, Germany

www.ingramcontent.com/pod-product-compliance
Ingram Content Group UK Ltd.
Pitfield, Milton Keynes, MK11 3LW, UK
UKHW050229200926
475702UK00020B/2943
* 9 7 8 1 0 4 1 1 3 0 5 4 3 *